THE HEAT'S ON

SHORT STORIES OF THE SOUTHWEST

WILLIAM H. TRAUGOTT

Wasteland Press
www.wastelandpress.net
Shelbyville, KY USA

The Heat's On:
Short Stories of the Southwest
by William H. Traugott

First Printing – December 2015
ISBN: 978-1-68111-080-6

Printed in the U.S.A.

0 1 2 3 4 5

TABLE OF CONTENTS

PREFACE v

THE GEOGRAPHY xii

1. *Hot Ground* 1

2. *Citrus Orchards* 11

3. *Devil's Highway* 21

4. *Campground* 29

5. *Campfire* 35

6. *Wind Machines* 46

7. *Backfire* 56

8. *Windows* 63

9. *Coal* 69

10. *Irrigation* 83

11. *Climate Change* 93

12. *Below the Dam* 104

13. *The Vehicle* 112

14. *The Big Trees* 118

15. *Ice Age* 128

16. *Highway 66* 133

17. *Canyon Country* 146

PREFACE

As climate change and global warming occur, continents are experiencing drastic shifts in weather and surface conditions. The Southwestern United States, one of those regions, is experiencing warming conditions that have created a drought with rising temperatures pervading from the West Coast inland while citizens feel the heat. *The Heat's On* is a significant expression that applies climatically as well as historically. When a furnace is turned on in a cold home, the householder announces, "*The Heat's On!*"

The Heat's On—polar icecaps are melting; glaciers are receding; oceans are rising; atolls are being swallowed up; tides are getting higher; fires are increasing; forests are burning while tornadoes are occurring year round; drought is rampant; and, besides all that, average temperatures are rising globally. The climate is changing, and regions are experiencing weather never before realized in recorded history.

The Heat's On—fossil fuels buried for millions of years are being brought to the earth's surface and are being burned to feed the modern industrial revolution. Carbons are being released into the air by the tons every minute. Recorded temperatures are rising by the year and by the decade.

The Heat's On—this collection begins with a story of a people whose name means "without water " (Latin, *sine agua*—Sinagua) and concludes with one that focuses on the river which provides water virtually to the entire Southwestern United States. Climate change is hitting the entire globe; but how it, together with drought and heat, affects a particular region—the Southwestern United States—is the focus of these stories. Many elements of the region are being changed by global warming.

Earth, air, fire, and *water* were the classical Four Elements in the ancient thinkers' analysis of the world. Considered seriously, the four are beneficent, necessary elements for life. Humanity lives on the *earth* that provides sustenance for living. *Air* contains oxygen—essential for human existence. *Fire* provides the means to convert elements into energy that warms human beings and drives the mechanisms that have made life possible. *Water* sustains life, for living things exist by and in it—human beings are more than sixty percent water. Seen in extremes, change that affects habitable regions comes with these elements:

- **Earth**—*Desert*—no water, thus, no life
  Polluted chemically with inability to sustain life
  Molten-volcanic-with transformation of surface and destruction of living organisms
  Industrial fertilizers replacing compost and organic substances

- **Air**—*Tornadoes, hurricanes*—increased velocity with added destruction
  Contaminated atmosphere with marginalized oxygen and excessive carbons endangering life
  Ozone holes increasing propensity to melting ice, skin cancers, and endangerment of life through ultra-violet radiation
- **Fire**—*Unharnessed Fire*—forests destroyed and photosynthesis compromised for conversion of carbon to oxygen
  Industrial pollution through burning carbon with its increase atmospherically from vehicles and factories
- **Water**—*Drought*—lack thereof with life shriveling up
  Polluted—stagnation, industrial waste, chemical poisoning
  Excessive—flooding, erosion, tsunamis
  Freezing—destroying hail, cold weather where not usually expected

The list could grow and be enhanced, but the delicate nature of life has been realized by both ancient and contemporary inhabitants of this globe in appreciating these "elements."

The extremes of the elements—both destructive and beneficent—have been referenced in religions' sacred literature:

- Experiencing life in and beyond time

- Eternal destruction with imagery of *heat* and *fire*
- Heaven with *water*, river (Revelation 22:1)
- Life's *waters*—the Lord "leads me beside still *waters*" (Psalm 23, Judeo/Christian sacred literature)
- Conquering harsh *wind* and high *waves* by deity
- Deserts blooming—a region of habitation by industrious Israelites and Christians

Judeo/Christian literature is filled with human beings' experiences with the elements in their support and destruction. Even in the contemporary world, the Israelis have continued to survive and thrive in the desert—instituting drip systems for irrigating plants in what had been only desert. Yet, evidence for climate change and earth warming mounts!

The Heat's On—worldwide heat records are being set. The hottest average temperatures globally for the month of May—since global records have been kept—occurred in 2014. Drought—evidence of the heat—was being felt in the month, especially in parts of the Midwestern and Southwestern United States. Glacial melting was apparent, especially because Alaska had one of its warmest Mays since climatologists started keeping records for the state. Meteorologists show pictures of the arctic melting and Greenland's losing its white, snow cover.

The Heat's On—and not on the cook stove, either. Citizens complain and tragically at times die of the heat. Air-conditioning is being installed in places that only a few years ago went without. The population shift in the United States to the South and Southwest occurred when air-cooling was technically possible and economically feasible. "What is this world coming to?" some old timers have asked over the generations. It's a pertinent question.

The Heat's On—in the waters of the world. Oceans' temperatures are rising. Coral reefs are dying, and fish are migrating to cooler waters. Melting icecaps mean less white surface to reflect the sun's rays back into space, and that creates more heat. The frozen fresh water is melting into the salty seas.

The Heat's On—drought continues to increase world-wide while some areas are experiencing flooding. With the drought conditions, a number of cities of the Southwest are considering desalination plants for generating fresh water and, in some instances, are putting them on line.

The Heat's On—more pollutants are in the air, and asthmatics are suffering increasingly because of what humanity's machines are belching into the air. The United States has led the world in owning automobiles, but Third World countries are now adding more vehicles. Increased carbon accelerates global warming. With less protection from ultraviolet light, skin cancer can be on the rise. States in the

South and Southwest particularly have high rates of carcinoma.

The Heat's On—clean energy is part of the answer. Fossil fuels have been the basis of energy for the Industrial Revolution, but clean energy sources—renewables—can reverse the trend. The solar resource has been researched and is being developed for its thermal energy, photovoltaic capability, and biofuel production. Flat-plate collectors gather the sun's heat that provides hot water, while photovoltaic cells generate electricity directly from the sun. Heliostats, set in arrays around a thermal tower, generate concentrated heat that melts salt that boils water to create steam to drive turbines for electricity.

The Heat's On—"Decision time has come—it's time for action," advocates are urging as pressure is being applied for something to be done about industrial pollution and an energy substitution through renewable energy. "The heat's on for humanity to do something about this," they say. Denial of the reality may be regional or sometimes national, but today's science presents the evidence.

The Heat's On—each of these short stories tells of the human experience regionally—the United States Southwest. What people's experiences have been; how they have coped with the situation; how the changing conditions have and do affect their lives—these are the stories' themes. A very common experience for Southwesterners is high temperature, and in excess it can be devastating. From the experiences of the early Sinagua tribe in Northern Arizona to contemporary citizens of the Southwest, these stories relate the human cost and effort to deal with regional change in climate and temperature.

With the delicate tolerance of the human body for extremes in heat and cold, testimonials of the human experience—both living and dying—come from the millennia. Some people's stories speak to extremes in: drought and famine; rain with flooding and drowning; tornadoes and willy-nilly destruction; fires and burning; hail and devastation; hurricanes and total annihilation. *The Heat's On* features fictionalized accounts of global warming so very real to the citizens of this land—their everyday experiences in the Southwest.

The Southwestern States, traditionally, have been: Colorado, New Mexico, Utah, Arizona, California, Nevada, and part of Texas. From these states comes this collection of stories about global warming—climate's effect on life in all its harshness and kindness, humanity's experience of struggle and triumph.

From one Southwestern citizen
and concerned traveler,
William H. Traugott, Th.D.

THE GEOGRAPHY

1. Hot Ground

Setting *Arizona: Sunset Crater and Wupatki National Monuments, Flagstaff*

2. Citrus Orchards

Setting *Arizona: Salt River, Roosevelt Dam*

3. Devil's Highway

Setting *Northwestern New Mexico: Highway 666*

4. Campground

Setting *Colorado: San Juan Mountains, Ouray*

5. Campfire

Setting *Arizona: Mogollon Rim, Phoenix*

6. Wind Machines

Setting *California: Palm Springs*

7. Backfire

Setting *New Mexico: Jemez Mountains*

8. Windows

Setting *Arizona: Urban Home*

9. Coal

Setting *Arizona: Navajo Nation, Kayenta*

10. Irrigation
Setting *California: El Centro*

11. Climate Change
Setting *Texas: Guadalupe Mountains, Small Town*

12. Below the Dam
Setting *Colorado: High Plains, River Country*

13. The Car
Setting *California*

14. The Big Trees
Setting *California: Sequoia National Park, San Joaquin Valley*

15. Ice Age
Setting *Southeastern Arizona: Interstate 10.*

16. Highway 66
Setting *Southwest: Highway 66*

17. Canyon Country
Setting *Colorado: Middle Park and River Country*

1

Hot Ground

Setting Arizona: Sunset Crater National Monument, Wupatki National Monument, Flagstaff—eruption of Sunset Crater (11th century) and destruction of the Sinagua village, whose inhabitants may have moved north to Wupatki

Hot Ground—that's what the woman felt as steam was rising from fissures in the earth. An afternoon breeze wafted through the tall ponderosa pines that surrounded the tribe's pit houses, but the mist coming from that crack was hot. With her companion, she had experienced mist and rain falling on their home, and snow drifting down in the winter. However, never had she experienced hot mist from the earth. Shaking had awakened the village several weeks before she saw the first break in the ground with the rising mist. What was happening to her and the tribe?

As she stood there looking at the trees and down at the fissure, she remembered several weeks before, when she had been startled with the ground quaking. Just as the sun was coming up, her sleeping mat had been shaken violently—never had she felt anything like that! The rumbling had even dislodged some stones which had been stacked as a wall for one side of the pit house, where she and her companion lived. The village dwellings were protected with roofs of supporting poles covered with pine boughs, but on that day they did not all stand. Cracks were opening up in the ground. The loosened stones of her house had even moved a couple of supporting poles just above her. Repairing that wall and roof had required some work by the men, who ordinarily were out hunting for game. The earth had settled down after that first quake, but then a few days later several cracks in the earth started to open right where they were standing in the village. The hot mist had started coming from a couple of the cracks. Those crevices had broken the surface in several places. One of the fissures had even opened right across their gardening area.

This morning she had been awakened with her sleeping mat moving—again! Startled, she had gone outside to discover that one of those cracks had moved further through a couple of the mounds of plants that she had tended so faithfully. And then that hot mist, steam, really started rising more intensely from the fissure. In her mind, on this breezy day, she thought about what had happened—the quaking had preceded the cracking, and then the steam had started coming from the fissures. What next?

"How dare that crack!" she thought. It had worked its way across her garden! She was not happy about that—in fact, a certain fear of the unknown overtook her. Then the steam had really started pouring from the fissure! With the other women, she had kept mounds of beans, squash and gourds that were faithfully watered from a small pond since early in spring. That small source of water was the result of generations of hard work—a dam built to retain the meager moisture that came to the area. In fact, the precipitation was so scarce that later Europeans arriving on the scene called her people the Sinagua tribes—natives "*without water.*"

As the woman stood there clearly fearing the changes, she wondered if the gods were angry with the tribe. Their shaman was challenged to explain what was happening to the tribe. He had told stories of their finding this place many moons ago. Coming from a warmer climate, their ancestors had found the high ground, cool days, tall trees, and plenty of animals for hunting. The men trapped rabbits and with spears in a big hunt sometimes would bring down an elk. Both large and small animals—elk and rabbits—had become part of their diet. With their spears, the men quietly tracked the large animals—downwind so that the herd did not sense their presence. Walking stealthily across the pine boughs and needles, they learned over the years that one crack of a small branch could startle the herd—thus, the reason for caution. Using every part of the animal, they cured the meat and tanned the hides that provided them warm garments for cold winter days.

On this day, the woman stood there trembling. Over the next days she felt more quaking. The cracks grew in size, and one finally went right through the dam which held back that pond. On that afternoon, the water flooded the village, and what remained of the mounds of vegetables was washed out. She was terrified when the flood of precious water came tearing through their small village. The tall trees shook but stood firm while everything in sight washed away. All the generations of work on the dam—their hard labor this past spring to plant seeds and nourish the precious plants with water and good soil. All gone! What happened? What had they done to displease those gods?

Surviving the flood and still clinging to a large pine tree with her mate, she heard a huge roar in the distance, while the quaking increased. They could take it no more! Running in terror and leaving the momentary protection of the big tree, they fled with the other tribesmen toward the setting sun, across an open meadow—taking no time to look back. Fine black soot and flakes started to fill the air and land on rocks around them. Finally they reached higher ground once more with large trees and boulders around them. Suddenly they heard "plops" on the open plain that they had just crossed. Smoking black stones were landing on the ground. They found an overhanging cliff with a small cave and huddled with other survivors of the cataclysm. The trembling couple finally had the courage to glance over their shoulders—the devastated area from which they had just come was covered with black soot and smoke. They could see streaks of fire coming from a hole in the midst of the cataclysm. The earth now tembled more violently, and hot stones were being thrown higher and higher into the air. Protected by the overhanging cliff, they pushed themselves deeper into the back of the small cave.

As they trembled and held one another, the couple pushed themselves against the rock as though they were trying to make a new impression on their faces. So terrified, they could not stop trembling. The two had little left—the loin clothes that they wore and some skins over their shoulders. She had managed to grab a leather bag of dried maize, while her companion had whisked up a large woven basket containing additional seeds and a couple of gourds, one with some water. They dropped these few belongings in the back of their rock enclosure. The ground under them continued to shake, and they feared for their survival. Dust fell from the top of their small cave, but at least they were not being pelted by falling stones as were other fleeing survivors.

The acrid air held odors of what smelled like rotten birds' eggs that had fallen from old nests and cracked open. In their sanctuary, they were at a distance from the steam vents. The falling ash mixed with the hot stones did not reach the very interior of their enclosure. Periodically they felt wafts of hot air, however, blow over their sweating bodies. Survivors' moans of pain were heard, and the recurring stench of burning flesh drifted into their enclosure. The couple's years of endurance of cold and heat, injury and disease had hardened them to suffering. The tribesmen uttered no screams even as they endured excruciating burns on their exposed bodies.

From their momentary sanctuary, the couple finally turned to study what was happening outside. Little could be seen of the sky itself, for the sun was blocked by the smoke and ash. In the distance, a yellow and red spewing was visible on the other side of a small hill that protected their small cave from the erupting cataclysm. A yellow glow was emitted over the entire scene, as the couple clung to one another and with other terrified tribesmen stared in amazement at reflections of destruction outside. A distant roar obliterated any small sounds around them.

Finally feeling some sense of security, they quietly checked one another for any burns and injury. No bones were broken, and she had only a slight sprain on her right ankle as she had stumbled over a protruding rock that the meadow's grass had concealed. The flying stones from the eruption were hot and hissed whenever they landed in a bush or patch of grass. Not having eaten or drunk anything through the eruption, the two began to experience hunger and thirst. They fumbled in the darkness of the small enclosure to find the few belongings that they had managed to grab in their flight. The gourd of water had not spilled, and they took a few sips of the precious liquid partially to quench their thirst. Then slowly chewing a few seeds, they returned to stare at the source of the yellow light that glowed nearby and illumined the walls of their enclosure. They were safe, and finally exhaustion drove them to sleep in one another's arms.

They awoke the next morning not to the sun but to the continuing glow of the distant eruption. The roar continued, but the falling stones had abated. Fear drove them to stay within their enclosure in the early hours of the new day. As they peered through the cloud of ash, they saw some bodies slowly being covered with a pall of black ash. Growing stiff in their secluded enclosure, they tried mid-morning to venture forth from their confines. However, finding it difficult to breathe outside, they returned to their small cave. They removed some of the rocks from the back of the cave and placed them in front as some protection from the chaos outside.

The precious water in their gourd was disappearing, even as they only took small sips to address their thirst. By partially enclosing themselves with that temporary wall in front of the cave, they managed to find some comfort and protection.

Although the roaring continued, the ash had begun to clear with a change in wind direction, and the stones no longer were dropping on the second day. In that lull, they decided to escape from their life-saving sanctuary, and to move into the wind that was blowing the ash away. They found a path and followed it northward through the shrinking forest. Eventually they could look back and see only the glow of the eruption against the surrounding hills. The two discovered other members of their tribe—including the shaman—emerging from the forest to which they originally had escaped. Fearing other tribes that lived to the north, the couple joined the survivors. With the medicine man's encouragement, they no longer feared the gods' wrath and thought about rebuilding in a grassy meadow safely away from that growing mound of hot rocks.

In the first weeks, they constructed rudimentary shelters with trunks and branches from nearby evergreens. Piling rocks on the side toward the rising sun, they were able to capture the warmth of the afternoon sun, for the intense heat from the eruption was at a distance now. Slowly and then more rapidly, a new mound formed around the fire and rock coming from the earth. While the women worked on their shelters, the men managed in those first days to find animals that had died in the initial eruption—they had some food to survive. Drying the meat, they created their own jerky for the coming winter.

The fierceness of the eruption abated, and having moved at that safe distance from the forming mountain, the survivors established a new village in the months that followed. In all the chaos, they had managed to save some of their precious seeds and in the following spring planted new gardens. Being on a trade route, travelers who soon found the tribe's new village shared by sign language and sketches in the sand what they had seen—distant smoke rising from the ground and a growing mound of rock. They described their fear to come closer to the eruption. Approaching the billowing smoke, they had heard the growing roar. The cloud above the emerging mountain with the spewing rock appeared to be a lightning storm with fierce thunder.

In the next years, the couple joined other tribal members and gathered flat rock of sandstone and basalt to construct their masonry walls; placing mortar between the stones, the builders sealed the rock together to create a solid wall. Thereby, they moved from their destroyed pit houses to a pueblo—joined rooms for greater security. To survive, the men hunted with their spears and found antelope, rabbits and other game that provided a mainstay for food. The new mounds of seeds and plants were small at first, but constructing some retainer dams again, they were able to bring water to their precious gardens. After several years, their initially meager walls grew into a pueblo with shared rooms. In the first year after their escape, the mountain grew, continued its eruptions, but over the years those eruptions lessened and finally died. The new village used some of the ash in the soil for their gardens and even had opportunity finally to build a kiva for their sacred gatherings.

To the couple, several children were born, and the small tribe grew in number. Their crops increased, and travelers brought more seeds, trinkets (including necklaces and arm bands), bird feathers, and other items not available to them. The mother joined other tribal women to weave baskets for their own harvesting and storage while trading some for goods brought by the visitors to their pueblo. With time, she added ceramic pots made from the clay around their new home. With the passing of many moons, this survivor saw her children grow up and construct new rooms for their partners and families. Around the evening campfire she often recalled the fire, smoke, and ash that had come from what now they saw as a high mountain. She drew the picture of how in its youth, the mountain had been fearful and angry—spewing out rock and ash. With time, the stories grew and were embellished, but having escaped, the tribe's members never again in their lifetimes had to fear the *hot ground.*

2

Citrus Orchards

Setting Arizona: Salt River, Roosevelt Dam. 1911

Citrus orchards—he was thinking of them, as he waited in the crowd of nearly one-thousand people. His former President was coming to dedicate the dam that finally would control the waters of the Salt River. The day was warm, even for March, and they had been waiting for a while as he stood on a promontory overlooking the dam and lake—fully visible from where he stood. A large crowd had gathered on this March 18, 1911—all in response to newspaper announcements regarding the event and the reason for the President's visit.

The Valley of the Sun had a flooding problem and a sporadic water supply, and a dam was required to control the situation. A dam—that's what they needed to water their orchards, and Valley residents had pushed for it for years. Its origin really went all the way back to 1902 with the need for water control in the Salt River Valley. Today they were dedicating the Theodore Roosevelt Dam, known to some as the Salt River Dam. Ralph mused over the past months and years—all those newspaper articles, the politics, the growers, the divergent groups, and the local and national debates. In his waiting at the moment, his mind raced over the events that led to this day.

* * * *

The man coming today is out of the *Presidency*, Ralph mused. President Theodore Roosevelt had come into office as a result of the assassination of his predecessor, President William McKinley in 1901. He served the remainder of that term and then was re-elected to a term in his own right, thus serving from 1901 to 1909. "William Howard Taft is in the office at the moment," Ralph was thinking, "but I hear that Roosevelt still has some interest in the office. It sure has sounded like he might try again." Immediately after his Presidency, Teddy (as he was known to many citizens) had left for a safari in Africa. Ralph also recalled reading that, upon Roosevelt's return, he was as popular as ever, went on a railroad tour of the country, and was cheered by thousands of citizens who really respected him. Ralph was in that crowd—both in that group of followers as well as in this crowd of well-wishers today.

* * * *

As he was recalling details from his readings, his eyes scanned the large group, and he was jolted from his daydreaming when he recognized an old acquaintenance in the crowd. Weaving through the gathering, Ralph found his friend, who was as surprised as Ralph to find a familiar face in this press of people. "My goodness, it's great to see you again" were Ralph's first words. An exchange of convivialities followed, and Ralph's curiosity led to a question. "What brings you here?" The friend quickly went on to explain his part in planning the day with other Valley notables. "Wow, then you are an authority on how this all got started!"

"Well, not exactly. I just heard that the President was coming and wanted to be a part of helping. Really, I only have heard a little bit about the struggle back in Washington that brought us to this day," the friend responded.

That was an opener for Ralph, always ready to move into a subject with detail, to start an explanation of what he had just read before coming to this celebration. It was his opportunity for a flashback to *the Act*. There had been some controversy, Ralph went on to explain, regarding providing funds to Arizona, not yet a state, under the Newlands Reclamation Act of 1902. The original Act was established to open up undeveloped lands of the West. Since there were orchards and canals in the Valley of the Sun already, some question had been raised about the availability of the new money, but that was cleared up, and the plans proceeded. The Act also established the United States Reclamation Service, which was to design and construct irrigation projects in the West to assist in settlement. By now the friend was intrigued, for the President hadn't yet arrived, and he liked hearing the actual story.

Ralph went on to explain how the construction of the dam was authorized under that Reclamation Act—one of five federal projects authorized at the time—and construction started in 1903. "Well, you know how individuals and private irrigation companies had tried unsuccessfully to establish major irrigation systems in the West," Ralph noted as his friend assented with a nod. "But it took the federal government to develop large-scale projects."

With that much detail and one-way conversation, the friend obviously was becoming a bit uneasy, made some excuse about looking for something to drink, and blended back into the crowd. A little puzzled about the departure, Ralph just shrugged his shoulders and returned to his place on the overlook.

* * * *

Still no President—Ralph decided that standing on that high promontory was about as good a place as any to keep track of the action. He gazed over the placid water to his east and then to the mountains beyond. His mind flashed back to the *dam's construction*—from natural rock or "rubble masonry"according to one source—and the floods that it would control. They had constructed this dam on a spot designated "The Crossing." Ranchers, farmers, and Indians had used the spot before the dam to ford the Salt River. The crossing was a short distance from the confluence of Tonto Creek with the Salt. With the construction starting in 1903, the dam was completed in 1911 and stood as the largest masonry dam in the world at that moment, while the lake behind it for a time was the world's largest artificial reservoir.

Being an avid reader with almost photographic memory, Ralph recalled facts from an article that he had read several years before—including some details about some dignitaries who had gathered after the first stone had been laid in 1906. According to the reference, the dam was to provide irrigation water for the Valley of the Sun and flood control as well as hydroelectric generation. Flooding had been such a destructive reality to growers in the Valley that a major purpose for the dam was to control the water when downpours hit the desert and barely vegetated mountains. And here he still stood with the crowd—waiting for the President, looking down the canyon below the dam, and imagining what it would have been like to watch the Salt River, in full flow, roaring through these rocky banks and spreading then across the Valley beyond as it came out of the narrow canyon.

* * * *

Ralph's sources had underlined the need for control of the Salt River. With no dam on the White Mountain Apache Tribal section of the Salt River, *flooding* had been common. Even during the dam's construction, floods continued to be a problem and had delayed some of the work. "The devastating floods of 1905," Ralph recalled in his continued musing, "had interrupted the construction, but engineers and contractors prevailed and resumed the work." Again he remembered, with more detail this time, "At completion, this dam's height—280 feet—with its length—723 feet— made it the largest of its kind in the world. Wow! And then there's the lake behind it—the largest of its kind in the world." That was the monster to hold back the flooding and as a by-product to generate electricity through its hydroelectric system. For this, Ralph and the other attendees were very grateful. Above the dam on the river, locals and eventually tourists could prevail upon the Apache Nation for permission to float down the river, but that was for another time, not today.

* * * *

By now it was early afternoon, and Ralph had brought some oranges with him for a snack. The President still had not arrived, but the crowd had grown. As he grabbed an orange from his sack, again, Ralph's mind went back in time. His father had told him about the orange trees that had been brought to the Valley in the 1890's from southern California. *Citrus fruit,* especially lemons, was in great demand to combat the scurvy common among Western miners. With earlier ripening in Arizona than California, the fruit was in demand. Hundreds of acres went into fruit trees—along both the Colorado River in Yuma and the Salt River in the Valley of the Sun. That's from where Ralph's orange came—his own orchard right here in the Valley. Irrigation canals had supplied river water to the orchards, but control of flow was a challenge, and the desert always needed water—especially during the extremely hot summers. Anyway, Ralph enjoyed his orange, and as he finished it, the ever-alert crowd spotted some dust coming up the canyon in the distance.

* * * *

Then, he caught site of *the car*—President Roosevelt indeed was coming. It was 4:15—he had been waiting since early morning and had watched the crowd grow. Suddenly guns were discharged—saluting the President—and the crowd went into an uproar with cheering. Later Ralph would read in a Phoenix newspaper that eleven guns were discharged that day at the dam. Teddy, as he was affectionately known to many, attracted people by the thousands wherever he had gone for speeches or special presentations—and today was no exception! "What a construction job! But would this really prevent the flooding? They had all kinds of trouble just putting this thing up," Ralph thought, but casting aside these momentary doubts, he joined in the cheering. They were over a thousand strong that day, as he and the crowd glanced over the imposing dam with that United States flag flying from a parapet. The President waved, left the car, and slowly walked through the admiring crowd to the rostrum.

* * * *

Applause from the crowd brought Ralph from his westward gaze back to the President. The speaker was saying something about this dam being a monument. As the President thanked the crowd for naming the dam and reservoir after him, he spoke of signing the National Reclamation Act and reminded the crowd that over half of his Rough Riders, who had fought in the Spanish-American War, had come from New Mexico and Arizona. Now he was giving them a fair deal, a motto from his Presidency. In retrospect, he spoke of two of the proudest achievements of his administration—the Reclamation Act that made this dam possible and the Panama Canal. Ralph, with the entire crowd, drank it all in and cheered again as the President payed tribute to the veterans and thanked the crowd for the honor.

* * * *

As the President was speaking, Ralph periodically glanced down the valley and saw the river serpentine through the gorge. Suddenly the speech was over, and the cheering died away. In a momentary silence—at exactly 5:48 P.M.—the President pressed a button that released the water from the reservoir. As Ralph watched the water gush from the outlet, he thought of the canals through which this water would flow. He was reminded of the ditches that had been constructed centuries before and the ones since re-dug and widened—it was a *long history.* Now these old waterways would have a new and adequate supply. Ralph remembered reading about how back in the 1860's the Hohokam canals were reconstructed—those had been dug centuries before the coming of the later Europeans. That reconstruction provided Salt River irrigation to produce crops for the cavalrymen and those miners that had come to the territory. As others made efforts to expand the system, the Arizona Canal—he remembered—was built in the first half of the 1880's. "Now, today, one push of a button by the President—that's all it took," Ralph thought as he watched the gushing outlet, "to bring water through all those canals and waterways."

* * * *

After all the festivities, the crowd reluctantly dispersed, for T.R.—as he was also known to many friends—today as always had attracted a large crowd and had kept his audience at rapt attention. He was an orator and known for long speeches, but that didn't matter on this occasion—the admirers just wanted to hang around. After the President's car left, Ralph joined the crowd as they started moving back to their vehicles. The return trip was dusty and slow, for he was behind a long line of traffic that crawled from this high point back to the city in the west. He was going back down the same dirt road that had brought him from his orchard. In another year Arizona would be admitted into the Union as the 48th State, and some day a part of this road would be paved. Today, however, he was slowly following the serpentine line of cars back home with the reassuring thought that this irrigation system should provide ample water for the growers and their *citrus orchards*.

3

Devil's Highway

<u>Setting</u> Northwestern New Mexico: Highway 666. Mid-Twentieth Century

Devil's Highway—that's what they call it! What a lonely stretch of highway to a stranger. He had been driving since leaving southwestern Colorado's San Juan Mountains and saw their route turn into treeless country. Snow-capped peaks gave way to flat plateaus and barren hills. Evergreens were replaced with periodic chico bushes—desert vegetation where it could survive. The evening sun muted the stark contrasts between some vegetated high plateaus and the tan, dry land below. Slump rock wedded the high plateaus with the lower flat, clay plain.

When would he get out of this? He and his wife had discussed a place to stay that night but never really settled on a town or motel. He just knew that they were leaving the mountains and had to get across this desert. No means in the middle of all this to make a reservation—telephones were back at the motel or in a filling-station somewhere ahead of them.

South-bound, he had seen the smooth mountain road deteriorate to a narrow, curving two-lane highway. He mentally complained of his lot in being stuck with driving this road—hot August, and the heat was getting to him. "Look at those beautiful desert colors." Jolted from his self-pity, his companion awakened him to the surroundings beyond the highway. Southbound and coming around a curve, he found a straightaway where he could glance in the direction of her pointing.

"Yes, it is lovely and very different from where we were last night." Hitting another bump in the road, his thoughts returned once again to his reverie. "Wow, how could anyone survive here? Living in this for generations," he mused, "would harden anyone to a life of survival." The afternoon sun was coloring the white vista with yellow hues while darkening the reds of the sedimentary rock on the distant buttes. Small, volcanic-sculpted peaks stood alone like strange haunting shapes casting shadows across the landscape. Returning consciously to his companion once more, he asked, "What did that motel attendant call this road this morning, when he heard we were headed south to that town down there on the Interstate?"

"The Devil's Highway," was the brief response.

"How right was *that*? What in the world possessed them to choose that name?" he asked.

"From 666," she responded.

"What's that got to do with the name?" he retorted, by now irritated by the winding road. This two-lane stretch was uneven, curvy with periodic potholes, especially along the edge. The high-crowned road had no paved shoulders, and erosion had taken out some of the support on the highway's edge. Bumps where rocks had pushed up tarvia under years of beating sun kept him awake, but even that didn't remove his weariness. It had been a long day.

"The Bible! Somewhere toward the end of the Bible there's a reference to the number 666 being a symbol for the devil. That's the number the State gave this road, and the name came as a result."

Hitting another patch of rough pavement, he frustratingly retorted, "No wonder they've called this the Devil's Highway. It's hellacious!"

The highway conditions didn't resolve the problem for that night. Where would they stay? To a very tired Lloyd, that issue really started to be preoccupying. After all, that would get him off this road! "A motel for tonight!" As the thought crossed his mind for an instant, a sharp turn and another pothole brought him to reality again. His traveling companion had been admiring the landscape. As in the past, once again, Laura could bring him back from his habit of self-pitying. "Look at that unusual rock!"

At that moment, he rounded another curve and found another straight stretch of this (for the moment) southbound highway. Glancing to the west, he saw a high pile of rock, and with a closer glance he saw, in what was now a setting sun, an eroded mountain that looked like a ship. Always the geographical guide, Laura exclaimed, "It's Ship Rock. That's a major landmark for the local native Americans." A lot of stories surrounded that rock—an eroded specimen of ancient lava protruding fifteen-hundred feet into the air.

"Yes, look at that image in the evening sun! It looks like it's just sailing through the shadows below."

Bam! Another pothole and back in thought to his situation. Changing the subject, he frustratingly turned to Laura, "This road is supposed to take us to a major Interstate highway!" That route carried travelers eastward to a metroplex and distantly westward through more reservation country. That setting sun kept reminding him of those night accommodations. They really hadn't known how far they would go, when they started this day back in Telluride. That phone (far back at the hotel) should have been used for a reservation, and now only an unknown filling-station somewhere up ahead was the next place to stop for a phone.

Yet, even that couldn't avoid reminding him of how tired he was. They had been switching as drivers, and this was his stretch. They were headed on what seemed an interminable distance through the afternoon and into the night to the next town at the end of this highway and on the Interstate where they might stay. Still a very real dilemma—and as he mused and tried to make time on the highway, they came around a sharp curve and a sudden real decline toward an arroyo.

There he was! A pickup with headlights illuminating rushing water over the dip in the highway in the early evening shadows. There had been no rain all day—only dark clouds to the north. Pitch-black in the arroyo, all Lloyd saw were the two headlights and a seemingly narrow stream of water. Coming around the sharp bend, his headlights finally highlighted the swift-moving flood. An instant later at his speed, he was on the water's edge, stepped on the accelerator, and plowed through the stream. His headlights momentarily were submerged, and water went everywhere—over and on either side of them. The momentum carried them through the stream, and they started to emerge on the other side. With a quick roll-down of her window, Laura could hear running water beside the road.

From where had it all come? Obviously a downpour in the mountains had brought a torrent of rushing water down this arroyo and across the dip in the road. Without a bridge, travelers were subjects of the elements' whims, and this twosome became a pair of the victims—almost! "Now I know why that pickup was stopped!" Lloyd blurted as he quickly reflected on what they had experienced. That driver, obviously, had stopped because of the water and was not ready to plow through a rushing stream. "What if a boulder had been washed across the highway just as we came through? We would have hit that thing, disabled the car, and been downstream—who knows where?" The pitch of his voice clearly attested to Lloyd's nervous excitement after the experience.

Motel reservations had been the last thing on their minds as they went through that desert experience. Still hot outside, they—now more nervous and cautious—stayed on course. But they still needed a place to stay. After a few miles—as always on a clear night in the Southwest—one could see the aura of a town in the distance reflecting against the surrounding hills. They were coming out of the near-disaster on the desert.

They finally started thinking about getting off the road for the night. Unknown to them, however, was the town's events for this summer weekend. As they came into town, they saw a lot of pickups and old cars variously parked along the side of the highway. Occupants obviously were inside and outside those bright-lit restaurants and saloons. As the travelers drew nearer, those occupants were on the sidewalks and wandering between saloon entrances—one group of four abreast with arms over their shoulders staggering down the walkway.

"Where will we stay in this town?" Lloyd nervously thought. What they didn't know was that this was the weekend when the native American tribes of the area had their annual pow-wow in town. They brought their handiwork, artwork, and set up booths for the hundreds of visitors who came to shop and buy. These tribal members celebrated and unwound from the ardor of their lives. "Relieved from the sameness of their everyday lives, they can unwind and celebrate," Lloyd thought as he glanced over the scene. Not part of the culture, Lloyd and Laura, however, were nervous. Trying a couple of motels, they obviously and quickly found that there were no empty rooms for this night. "Let's stop and rest in a parking lot just to get off the road," Laura wearily suggested.

"Nothing doing!" Lloyd quickly retorted. With thoughts of what happened to strangers in parked cars along American highways, he was concerned for their safety. Besides, stories of inebriated celebrants roamed through his mind, as he lamented the danger of guns in the wrong hands. Not carrying any firearms, he and Laura weren't about to be victims of anyone's trying to wrest a weapon from them or use his own to harm them. Beside the point—they needed a place to stay.

Without a further word to Laura, Lloyd went through the yellow-blinking light at the one major intersection and followed the highway out of town. He was on his way—where he didn't know, but certainly "out of town." "Let's hit the Interstate and find our way to the metroplex to our east!" Lloyd frustratingly informed his other half.

Laura clearly got the picture of Lloyd's attitude—low-surface anxiety and frustration—about the situation. "I'll drive us in!" By now adrenalin had sublimated his weariness, and he was driving on reserve juice. At this hour the highway was relatively clear of traffic. As the dark hours marched on, both driver and passenger were getting sleepy. Word games this time of the night just did not work to keep them awake; and without a reservation, they both quietly hoped to find a sign along the highway that would advertise a motel.

Well, their AAA travel book provided some help, and they had some ideas. As they neared the metro area, the aura reflected the ten-thousand-foot peaks east of the city. They were relieved to be near some accommodations. Fortunate for them, one of the Interstate signs featuring accommodations came up with the very motel that their travel book advertised. A quick, but weary, exit from the Interstate, and they saw the motel's lighted sign in the distance. A rather rapid swing into the parking lot, and Lloyd was out of the car, into the building, at the check-in desk, and conversing with the night attendant. Yes, they had a room, and without hesitation this weary traveler took the room and went to the car to bring in a very tired wife. The night was warm, and thoughts of the near disaster in the flooded arroyo still hung over them. Tomorrow Lloyd would raise their car's hood and find the sand caked on the engine and under the hood, but tonight they would enjoy the sleep of the "dead-tired," for no demonic plot had denied their survival of the *Devil's Highway*.

4

Campground

Setting *Colorado: San Juan Mountains, Ouray. Twentieth Century*

Campground—what a relief to be here! They had started out in Denver early in the morning, driven all day, and made it to the campsite just as the sun was setting across the valley. It had been unseasonably hot in the metro area—what a thrill to be at 9,000 feet in cool evergreens. The family had come from the Mile-High City across a number of mountain passes. They had reached Ouray at an elevation of nearly 8,000 feet and had seen that their camp site for the night was less than a mile out of town. Driving that last mile to the campground had seemed as though it were straight up—a thousand feet above the valley floor. Anyway, they had made it. Finding a campsite, father thought to himself, "Now to move and get supper started and the tents up." They didn't take long to get out of the stationwagon once it had stopped.

Being one of the organizers, father moved rapidly to unload gear while mom worked on finding pots and pans for the two-burner propane stove. Locating the tabletop stove, she headed for the picnic table, for the two parents and three kids were all hungry and tired. Having done this so often in the past, they were getting to be old pros. The kids each had unloading duties that they willingly took to—this time especially—because they were very hungry.

Not taking time really to look to the west, the family routine was interrupted when Josie exclaimed to her husband, "Heah, Jim, look at that sunset." The trees on the mountain across the valley were darkly outlined like sawtooth blades with the sun's orb still half visible and yellow beams of light radiating to paint the clouds above them with pink, red, and orange.

"Oh, my, that is beautiful!" Herb, the oldest child, responded. Jim glanced to where his spouse was pointing, but was so busy with unloading the remaining gear from the stationwagon's rack that he just smiled as their oldest son together with the other two siblings watched the sunset. The kids then proceeded with setting up the three tents—one for the parents, one for the boys, and one for their sister. Meanwhile, dad was too concerned about getting the campsite set up before it got dark to really take a moment to stare at a sunset. After all, they had to find the "facilities" in addition to setting up camp and fixing supper. He continued moving around the campsite—trying to help where he could.

Jim had felt a slight headache coming on when he first got out of their wagon, but didn't pay any attention to it—after all, "there was too much to do." Over the next few minutes, that slight pain went to a throbbing headache. He hadn't even thought of the heatstroke that he had experienced working as a contractor in the hot sun in New Mexico a couple of summers before. At the time, a couple of his carpenters had given him water, cooled him down, and decided finally to take him to the local hospital's emergency room. He had recovered with an overnight stay in the hospital, but the doctor warned him, "Take it easy in heat and at high altitudes. You could have a stroke from which you may not recover as easily as you did this time!" "Yeah," he had thought to himself, "I'll be careful but I do have work and a family!" That was then, but this is now—at a high altitude.

"Heah, Dad, what's wrong?" Hertha, the youngest of their three kids, asked as her father abruptly sat down on the nearest boulder. Josie, hearing their daughter's concern, immediately came over with a worried look on her face to check on her husband.

"Oh, I'll be all right," he responded with short breaths. "Just a slight headache!"

"Yeah, you know what the doctor said. You had better take it easy," came the response from the concerned wife. "Tell me what's really wrong!"

With that demand, he explained the headache and feeling some shortness of breath. "Okay, I know when I'm ganged up on!" He did rest, for the increasing pain made his head feel like it was being squeezed in a vice. Everybody gathered around: "I'll be all right. Just let me rest." So, with some hesitancy the family went about finishing the unloading and setting up the camp.

In a few minutes, Jim made a move to get up from the boulder and cringed with pain. "That was a feeble effort," he thought, "but there's supper to help with." He had to sit down again on that boulder. "So much for all that male bravado," he apologetically thought as he watched his family arrange the campsite. With a few moments rest, the headache began to let up, and the father of this operation thought he might move around a little. Any sudden jerks, however, brought shots of pain through his head; so his movements were purposeful and deliberate.

"How about letting me help with supper?" he volunteered, as he looked at his busy spouse. "I can set the table while you cook the hotdogs and a pot of beans."

"Are you sure you're able to do that?" his concerned spouse asked. "You'd better let me do this whole meal!"

"Oh, okay," he said reluctantly, "but I'm sure that I'll be all right. I'll just take it slower than usual." And indeed he did, for usually he was a dynamo around a campsite. Wincing every once in a while and hoping that no one noticed, Jim gave his other half, Josie, a quick hand grasp of affection as her hand passed over the table where he was setting things up. Herb, the oldest kid, stepped forward to organize the final setting up of the tents while Harry and Hertha hauled in the remaining gear. Their mother had finished final assembly on the stove, had turned on a burner, and then lit it as the propane hissed from the bottle. "How many does everyone want?" Josie asked as she pulled the hotdogs from the cooler. The crew came up with an order of ten wieners—hungry from the trip and at this high altitude.

With another burner going and a small pan to hold the hot dogs, the cook put a big pot of beans on the stove. Accustomed to family camping, everybody grabbed the necessary utensils, cans of drinks, and paper plates. With the food hot and ready, a prayer of thanks focused everyone on the beauty of the day and the warmth of family. The headache slowed down as its generator—dad—slowed down. Some deep breathing to increase the oxygen and some warm glances from the family brought Jim's focus to the beauty of the moment and away from that prior distraction. Tomorrow they would take the Million Dollar Highway south to Silverton and then to Durango. That heatstroke and the hot days back home were all lost tonight in the cool mountain air and anticipation of a restful night at this beautiful *campground.*

5

Campfire

Setting *Arizona: Mogollon Rim, Phoenix. Twenty-First Century*

Campfire—he just had to have one. It wasn't camping without warm flames licking around pine logs with hotdogs and marshmallows being cooked. "Sure, it's dry, and forest rangers have forbidden open campfires. But, by golly, this is my right! I'm a camper in a national forest! I'm out here to do some hunting, and *nobody* is going to stop me," he grumped to himself. After all, he had planned this trip several weeks ago, and he had *always* had a fire when he went camping. Traveling to a spot west of Snow Bowl, he was up here to escape the July heat in the Valley of the Sun. A news report he saw a night ago on his small, old television set featured a large fire that had grown to 10,000 acres in only two days in the eastern part of the forest. Tornadic spirals of flame had been created as the fire, pushed by thirty-mile-per-hour gusts, licked up the canyons and into fifty-foot, dried pines. The nation's largest single stand of ponderosa pines was under threat of destruction, again. The twelve-year drought had weakened the trees, and pine-beetles had finished the job. But he was ignoring all that since he was safe in his lair.

Before leaving his small house in the heart of the city, he had perused the junk under his portico and the remains of an old motorcycle and a busted rake buried in some prickly weeds in his backyard. He wanted to take a few tools with a long fork for his trip. He remembered leaving one in the backyard when he was trying to dig up a spot for a cactus he had bought. "After all, I need somethin' that's 'native'," he had thought when being "sold" on buying that desert plant. He scoured around the cactus—sticking himself with a thorn and getting madder by the minute. Finally, he found the fork and just slammed it into the back of the pickup to join the loosely strewn bolts and tools lying on the truck bed. However, he kept his precious shotgun and old pistol near his bed because he "never knew when sum-buddy might bother me." That "bother" meant wherever he might be. Putting his shotgun and pistol into the pickup's cab, he had left bright and early the next morning for the Rim.

* * * *

Hassen, Hass for short, had been born of a Hispanic mother and dominant Germanic father. Contacts later in his life had forgotten what his last name was—they just knew him as "Hass." His mother had loved him dearly, but his tyrannical old man had beaten him with a belt in his drunken stupors and whenever else he felt like doing it. As a result, he was growing up just as mad at life as his father, who finally one day just said, "You're so mad at life that your given name fits you—'*Hassen*', German for 'hate'. That's you—'Hass'!" And the nickname stuck. With that attitude, what would the future hold for this lad?

The small family had moved to the desert to get away from all the folks that hated his father, and needing work finally forced the old man to "come to town." In the meantime, the boy's mother had died a tragic death with cancer. With little guidance and a father that didn't care, Hass was in and out of school and finally at sixteen just plain quit. "I have better things to do," he told himself. By this time, he had experimented with drugs and alcohol, as a result of his father's addictions. Unable to support the habit and sickly to boot, the boy decided to move out and find work despite his physical condition. Bouncing between manual jobs, he worked on roofing, dug ditches for a contractor for a short time, and finally settled on carpentry. It paid, and he could come and go at construction jobs as he wished. Besides, the outdoor work seemed to strengthen him—especially now that he could afford to have three square meals a day. In his wandering around, he found that small, run-down house and left the ratty apartment that he had managed to rent when he left his hateful father.

Frequently unkempt, he never quite could be popular with the girls. So, he settled on being a loner and "meeting a girl" periodically when he was at a bar. Enjoying wandering, he came every once in a while to taking his beat-up pickup, which he somehow managed to finance, to the Rim. That Mogollon Rim was a popular place for second-home city-dwellers and for hangers-on like himself who just loved to be "up there out of the heat." Since hundred-plus days went well into October now that the monsoons were not around as much as in past years, many Valley residents enjoyed being on the cool Rim well into the end of October or beginning of November.

The Mogollon Rim provided an escape from the city's problems, and was beautiful this time of the year. The name had come from an 18th century Spanish governor in New Mexico. In late fall the few aspen groves in the forest turned yellow, and many campers loved to be "in the trees" and "see the color" that time of the year.

* * * *

Tonight the setting sun turned the sandstone and limestone cliffs into yellow walls framed by tall green pine. Virtually ignoring the constantly changing beauty, Hass was here on one mission—to have a fire and kill something to eat. The canyons had deer and javelina as bigger game, and there were always some rabbits or other little varmints around. It didn't matter whether it was hunting season or not to Hass; after all, he lived in Arizona where "many wanted to be free and not restricted by laws," as he wrongly thought. He didn't give a rip about those Feds, especially the Forest Rangers, who—with many loose gun toters around—had become more law-enforcement officers than interpreters of nature.

Hass had gotten to his camp site late in the afternoon—full well realizing that a fire was roaring many miles east of where he was. He knew about this place—it was down a rocky, rutted forest road, off the beaten path, where few people came. Away from everything—isolation, that's what he wanted both in the city and the country. Here he could do as he "dang well pleased." "Nuthin's gonna stop me," he often thought. Where would this care-less attitude take him?

He had a little daylight yet when he went gathering some dead wood from the forest floor and dropped the branches in a fire circle remaining from a number of other trips. With a couple of safety matches, he lighted some kindling that soon set the branches afire and roaring. For the moment he didn't pay any attention to the crackling sparks flying from the dry wood. He went to his pickup to unload his bedroll and a cooler in which he had fruit, hotdogs, and beers. Seeing the contents, he immediately reached into that cooler, grabbed one of the cans, and started guzzling a beer. A couple later and he would be oblivious of what he really had done. He threw his bedroll against a tree and returned to the roaring fire. "That's really goin' good!" he thought. But then he noticed some smouldering pine needles a couple of feet from the fire. His campsite was in an open area with only low ponderosa pines. Workers obviously had come through and cleared out underbrush, but the smoke from those needles was growing. The dry, popping branches had sent sparks in all directions. Already feeling the effect of the beer in his tired state, he was moving slowly until he realized what those sparks were doing.

Small plumes of smoke were coming from the needles all around the fire now and already at some distance. What to do? He ran for his truck and grabbed a towel off the front seat. Now moving in panic mode, Hass tried to smother the plumes of smoke, but there were more than he could handle. Meanwhile, that roaring campfire grew into a blaze a couple of feet high. Seeing the sparks spreading farther, he ran to the fire and started to kick at the burning branches—only, of course, spreading the fire more. Finally he thought to wet the towel in the melting ice of his cooler, but that took a moment after he finally turned to open the cooler and soak the towel. Slamming the towel against anything burning around him, he managed to douse some of the small smoke plumes, but the fire now was spreading.

Panicked, Hass could think of only one thing—what he had always done when things got too overwhelming—escape. This time he couldn't run—he could drive and drive FAST! Leaving everything behind—including a growing fire—he jumped into the pickup with his two weapons yet in the cab, fumbled to find the ignition with his key, revved the starter until the motor turned over, ground the truck into gear, and did a fast one-eighty to turn the pickup around toward the rocky road. It didn't take long for him to be speeding down the narrow forest road and bottoming out his pickup on some of those buried boulders in the road. The motor kept going, the transmission wasn't damaged, the axils held, and the tires didn't blow on the sharp stones, as he roared down the road. Coming to a sharp turn, he almost didn't see someone on the road just around that corner in the fading light.

Moving so fast down that back road and so determined to get out of there, he narrowly missed hitting a couple of hikers along the way. He braked and swerved just in time to miss them, but they didn't miss seeing him and his license plate. He floored the accelerator once more and sped on down the road to get out of there as fast as possible.

"It just got out of control. I couldn't help it!" he kept telling himself. He made it to the highway and was able to make time from the Rim. A couple of hours later he pulled into his Valley driveway. Trying to act nonchalant, he moseyed into the house from his carport without slamming the pickup door. Really shaken, Hass immediately went to the frig and grabbed a couple of beers. Guzzling the first one, he slowed down for the second and flipped on the television set. "Nothing new. Great!" he thought.

That night he slept nervously and woke up late the next morning to a monstrous headache. Quickly remembering the events of the previous day, he decided to take his pickup to a car wash to remove the mountain grit. "That should make it look better in case anyone really gets suspicious," he thought. Before he left, he popped a beer for his breakfast. Then he took off and found a filling station where he got gas and used its carwash for a quick run-through. Trying to lay low for the day, he went home, flopped onto the couch, grabbed a couple more beers, and turned on television. Emotionally drained, he quickly fell asleep with the television blaring.

Hass finally slept off the hangover and woke up just as he was hearing the evening news come on. He got up to go to the kitchen, for he was hungry and hadn't really eaten anything all day. As the news report played in the background, he heard something about another fire. The announcer reported "a small acreage fire just west of Snow Bowl in the Coconino Forest." Fortunately some hikers, according to the report, had spotted the fire and were able to contact Forest Rangers, who brought in a fire crew. "With the quick action of the hikers, another potential disastrous fire was averted," the reporter stated. "The one-acre fire was doused before it reached the taller ponderosas." That report calmed Hass somewhat, but he still needed some food and a couple more beers to drown his anxiety. That night he slept more soundly than before, but he still had to drown his worry in more beer. "Did anyone really see me? What of those hikers on the road?" he thought—finally falling asleep.

Hass woke the next morning to a hard knock on his front door. Headachy for a second morning from the beers, he had another hangover that made him groggy. He had difficulty waking from the whole previous day's experience, but the knocking now turned to pounding with someone calling his name. "Who in the world could this be?" He managed to slip on his tattered, stained jeans and stagger to the door. "Who's there? Go away!" he screamed. That didn't stop the knocking, for it resumed. Afraid to answer but unable to run, he hesitated and after more pounding he opened the door. "Are you Hass Hemet?" the voice demanded.

"Yeah, what's it to you?"

"Your pickup with that Arizona plate parked out there in the carport was reported driving fast from the origin of the Bowl Fire two days ago." By now, the small blaze had an official name, Bowl Fire.

"Were you on the Mogollon Rim day before yesterday? "

Now realizing that he could be in some trouble, he didn't answer right away and was asked the same question again. "Were you on the Rim day before yesterday camping west of Snow Bowl?" Now he knew that they had him.

"Well, no, I was here."

"Witnesses saw your pickup and identified the license plate. You own the truck parked there in that carport?" The two law enforcement officers had him, and his only answer had to be a hesitant, "Yeah."

"You're going to have to come with us."

"Lemme git some clothes on, okay?" He was nervous and shaking but still had some of that caustic belligerence in his voice. Going back to his bed, he slipped on a shirt and put on his shoes. By now the officers had come into the dirty front room wreaking of alcohol and waited. Hass was led to the squad car, taken to the local police station, and marched into a small room with two chairs and a table. Questioning brought a final admission from him. The witnesses had roughly identified him as he had sped past and remembered the plate numbers. As a result, Hass was booked for starting a fire in a national forest without a permit and leaving the scene.

After several days in jail, he was arraigned, charged with starting a fire, and bail was set. Not being able to put up the bail, Hass sat in jail—without one visitor. Assigned a public defender, he explained his childhood of fear and beatings; and admitting his guilt, he agreed to a plea for mercy from the court. Gone was the earlier belligerence, and with the defender's help, he put forward a plea bargain.

Finally, in court, before the judge, his defender made a point of the fact that Hass had not been arrested previously. "Your honor, you will note that in the accused's statement he had not maliciously set the fire—only panicked when he realized what was happening." The defender went on to emphasize Hass' childhood and running when fearful.

That plea touched the judge, and Hass was let off with a twelve-month probation, sixty days of public service, and a warning never to return to that forest again. "You are being released on probation for twelve months with the warning that you are never to return to that place for hunting or camping. In addition, you will go to a rehabilitation center for recovery from your alcoholism. If you are ever caught in the next year setting a campfire or hunting under these circumstances in that forest, you will be arrested and jailed." In the consultation prior to the court appearance, the judge had learned of the beer cans in the pickup and around Hass' home and had received reports from neighbors who testified to his inebriation on several occasions. That sentence put fear into Hass, and with that shock he nervously welcomed release from jail and the instructions on what the future held for him. Through his contacts in rehab, he made some friends, even a young woman with whom he found great affinity. With opportunity to consider what might have occurred, Hass changed his attitude and vowed that certainly in that forest he had started his **last** *campfire.*

6

Wind Machines

<u>Setting</u> *California: Palm Springs. Twenty-First Century*

Wind Machines on San Gorgonio Pass—he had heard about them. Why there? Wind! That was the answer! Having gotten his money in real estate, he made some queries about investments, and wind energy was a targeted field. "I want to put my money into a new and growing field," he had thought one day as he was considering retirement from his real estate empire. "Where could that safely be?" he had asked himself almost every day for the past month, since he had announced his early retirement to his staff. "I'll keep the business but turn management over to you," he had told his two managers. "That way you'll have security in knowing that our real estate operation will continue."

Carl's administrative assistant had set up a week at a resort in Palm Springs for him and his wife. In giving directions as almost an afterthought, Carl had asked his assistant, "Could you also call my Uncle Bob about getting together for an evening over dinner? His phone number is on my private computer file of VIP's." Well, all that had been done, and as a result Carl and wife were looking forward to this trip that was part of their "bucket list."

His office building in Indianapolis was a long way from Southern California, but he had decided to take a "driving" trip with his wife from their airline destination, Los Angeles, to Palm Springs, east of the city. Driving from LAX through the metro traffic was a challenge—even for this guy from Indianapolis. His wife Jeannine had wanted a convertible. "After all, we are in Southern California!" Carl was happy to do that, for he was enjoying the freedom and it was spring.

However, a couple of traffic jams over ten lanes in one direction on this LA freeway made his Mid-Western freeway experiences almost pale into insignificance. After all, this was the city once noted for its number of cars per capita and also its concentration of smog. That all changed, however, with stricter environmental state laws over the years and a great cognizance of cleaner air by the civic and state leaders—now he was headed for those wind farms. Well, almost, for they needed a night free of travel. As the traffic let up a bit and they found I-10 east to Palm Springs, they climbed out of the California basin.

Carl had heard a lot about the fine resorts in that desert area. Their plans took them across the very windy San Gorgonio Pass, as it was known there. Now they were enjoying the view from their convertible. On that drive over the Pass, he was looking forward to seeing all those "windmills," although that wasn't what his Uncle Bob had called them. "They're wind *machines*," his dad's brother had emphasized in an earlier telephone conversation. His uncle, who had worked as a researcher at a federal laboratory in suburban Denver for a number of years, had explained that these wind machines had been part of his research. Being Carl, he was ready to jump into investment in this "latest and greatest" means of making money and helping the environment, to boot. However, that meeting with his uncle was to be both relaxing with reminiscing and informative with the advice that this favorite relative would give—but that was still to come.

In the meantime, Carl and Jeannine were enjoying their trip from the coast into the mountains and the desert beyond. Los Angeles with all its vegetation and the ocean stood in stark contrast to the creamy white sands of the desert around Palm Springs. The Pass nestled between mountains towering over 10,000 feet above the desert below. With the open view from their convertible, the couple took in all the sites, and when they saw the first wind machine, all Carl could say was, "Look at that! There's one!" That first one grew into dozens—all in a row with fields of them standing starkly tall and stately above the desert floor.

A couple of highway signs warned them that they were approaching Palm Springs. Carl took the first exit off the Interstate that brought them past a large field of the machines on their left and the towering mountain to their right. The couple found their way to the entrance of a luxurious resort. The grounds were well manicured, and the portico at the entrance was spacious. Impressed and anxious to be free, they both welcomed the staff's attention and the chance to relax. After registration they learned of the resort's two, eighteen-hole golf courses, swimming pools, and lake. Carl thought, "Wow! What a way to let go and look at my future." This pre-retirement trip was the stage for a break and a future business investment—relaxing and profitable for Carl *and* Jeannine.

The real-estate executive soon was enjoying the visit and the chance to cast aside some of his Indiana obligations. He was quick to contact his desert relative. "*Uncle* Bob," he emphasized in the call, "this is your nephew Carl."

"Well, hi! How was your trip?" Bob responded with his usual conviviality. "When did you arrive?" In the usual familiar exchanges, they agreed on a time for lunch the next day with a tour of the area.

Concerned about their future—especially Carl's winding down from the business world—Jeannine was listening carefully and advising gingerly. She knew that her executive husband had been accustomed to "call the shots" in his office, but their relationship had been on a much more equal basis. They worked at their partnership, and both reassured one another that this trip was for a current break as well as plans for the future. Jeannine knew that her husband could not hang it up completely and would be looking for something into which he could sink his energy. That evening, they enjoyed an after-dinner swim and then a good night's rest. Both were anticipating the luncheon and afternoon the next day with some trepidation but also with confidence that they were on the right track for their transition and future.

Bob came by the resort and met the couple in the lobby. "Uncle Bob! The years haven't changed you," Carl complimented, as Bob approached them.

"I must say the same of you two. My, it's wonderful to see you. And, incidentally, at our age we can skip the 'uncle'. Let's just make it 'Bob'. My, I have looked forward to this visit." Bob exuded his usual camaraderie with hugs and directed the couple to his waiting silver Cadillac. There was no convertible for Carl on this day, since the local temperatures would be very high, and the sun's intensity would be just too much for the afternoon. The trip to the restaurant and the luncheon went smoothly. "So, how about a tour of the area and a look at the wind farms around here?" With that invitation, Bob opened up the afternoon's activity.

"Great! We are so ready and appreciate your kindness in taking the time and sharing your expertise!" On this afternoon's visit, Carl and Jeannine were to learn that many of these machines were being imported from Sweden, the major world manufacturer for the moment. But he also would learn that some new entrepreneurs in the United States were starting small with building their own blades and machines. Now that took space, and what better person to be in the business of finding space, people, and machinery for wind machines than a real-estate tycoon from Indiana! He had plenty of experience with space available from the failure of so much manufacturing in the Mid-West. Besides, he could afford to put some money into it. Carl started to share some of his enthusiasm about the possibilities.

"But be careful," Bob, who had retired to Palm Springs several years before, cautioned Carl. "Investigate buying shares in one of these farms, first, and then look into investment in development." Always empathetic to his uncle's advice, Carl thought to himself at the time, "Yes, that's good counsel. First I'll look into investing in wind farms."

The tour of wind machines went well, and concluding the drive, Bob added, "You undoubtedly have seen the tourist brochures on the area. I would advise your taking the aerial tramway up Mount San Jacinto. That will give you a panoramic view of the area—and you might even grab a bit of lunch while there. Hiking trails also go out from the visitors' center."

That sounded interesting to this environmentally conscious couple, for they wanted to do some hiking, if possible. The afternoon concluded with a drive through Joshua Tree National Park, a sweep past a field of wind machines, and a sunset arrival at the resort. Carl invited Bob for dinner, which the uncle graciously accepted. After the meal, Bob left the couple with the promise that he would provide recommendations on current and future investment possibilities. During the course of the day, Bob had shared his experience with the wind industry and his research back at the national lab in Colorado. So, now Carl could start developing his plans more concretely.

In his reverie that night, Carl remembered how he had started his business with selling homes, and then had seen that grow into selling office buildings. Soon he had acquired sufficient funds to do some buying of his own. After all, there were plenty of old homes around that needed fixing up and flipping. And then there were those old office buildings requiring renovation and refocusing. So, he had acquired some old warehouses in an area near some redeveloping neighborhoods. That had seemed to be a safe investment—and it certainly was. New college graduates had been looking for office space in their own real estate businesses, and doctors establishing new practices required sufficient space for installation of medical equipment. And they all required plenty of parking space.

All that selling and developing and flipping took a number of years, but Carl had the "golden touch," as his father had said. It seemed that whenever he tried something in real estate, he made it. He had a second sense on when to buy and sell. "Impossible," some of his friends had thought. It was real, however, for Carl. After all, he had membership in several civic groups and knew a number of bankers, lawyers, and politicians. He recalled how he and Jeannine had settled into a fine suburban home on several acres with plenty of room for entertainment. On this night in Palm Springs, he thought about all that warehouse and office space under his control and more that could be acquired—it would make great space for manufacturing wind machines.

Carl woke up the next morning early and started rapidly writing on his laptop. He was taking notes to himself and composing e-mails to staff. This visit to the resort was not all relaxation—Carl was in his executive, business mode again and planning fast. *However*, there was also the need to transfer some funds from his investment company into his bank account to start acquiring shares in wind farms. That meant contacting his certified public accountant—just "my CPA, Charlie," to Carl. This new investor in wind wanted some direction on investing in some of the companies that Bob had mentioned. By the time he had composed some of the terse, but key memos to his staff and CPA, it was well past sunup. Jeannine had left him alone, for she knew that he could not start this day without acting on the information that he had learned the day before.

The two decided to have another swim after breakfast and then to take a leisure trip up the tram. They had lunch at the top, hiked some of the trails, and cooled down with a drink at the bar before going back down to the car. Back at the resort, they enjoyed the evening with a stroll on the grounds and an early "retiring." Relaxed and enjoying their trip, the couple continued their visit with a couple rounds of golf, more dinners, and another visit with Bob. That visit became a consultation and Bob's confirmation of what Carl had already done with the provided advice. Again, Jeannine was in the background but very much a part of the planning.

The couple bid a momentary farewell to the resort and Palm Springs with the promise to themselves that they would soon return. Through Bob's advice, Carl had some new contacts to make with owners of wind farms and had to obtain more information on land investment in California. His mind was a swirl of details on what the future held, but Jeannine kept him grounded as he shared with her some of his plans and dreams.

This husband had a tendency to get ahead of himself; and, as usual, had to be slowed down. What better person than his life's companion? She reminded him that he would have to find someone who could ramrod the operation—technicians and entrepreneurs who would be willing to start the ball rolling. He would provide the money but, for employees, had hopes to go to a local college that had been training students in wind energy. So, he and Jeannine said "goodbye" to Gorgonio Pass—only for a brief time—and headed to LAX with mutual assurances that they had a "retiring" future in *wind machines*.

7

Backfire

Setting New Mexico: Jemez Mountains. Twenty-First Century

Backfire—not from a truck! They were out here setting up a backfire to burn away vegetation and to stop the advancing horrific firestorm. A bulldozer was plowing up ground to create a wide swath as a firebreak, while they endeavored to set a backfire also. Time was of the essence, but progress was slow. The mountain terrain was rugged. Trees grew between boulders and were lodged in crevices. The crewman managed to find a relatively level area, and that's where the dozer was doing its work. The wind—for the moment—although gusty was blowing against that catastrophic fire—thus, the right direction to set a backfire while they also tried to create a firebreak.

"Is this really going to help?" Mel thought as he worked the area. He was on the line with other firefighters digging with shovels and clearing underbrush, in anticipation of setting the backfire. The dozer, however, would have to complete its clearing while Mel and the other fighters followed to remove some of the scraggly underbrush. Hopefully, a sufficiently wide firebreak would slow the advance of the big one. But that nagging question kept returning to the firefighter, "Will this hold and stop it?"

That massive, catastrophic burn had all started when a crew of foresters had been contracted to thin out ponderosa pines, other evergreens, and underbrush high in the Jemez Mountains of New Mexico. A fire several years before on the local National Monument prompted foresters to plan this thinning and controlled burn. They started on the shoulder of a high peak west of Los Alamos. Mel had heard about the prior workers who had cleared underbrush and had thinned out some dry small trees that had sprung up from a wet season several years before. The recent crew of foresters was to have burned that underbrush and scrub trees that had been piled up for the right day. Well, the day that the crew picked wasn't the right one! Even with fire equipment around them, the flames got out of hand and licked up into the tops of the nearby dry pines. It went in a flash, and now Mel was standing in this row of firefighters east of that site trying to set a backfire and really wondering, "What were those guys thinking when they set that fire on a windy day with such terribly dry conditions?"

They were hearing reports about the eastward progress of the fire. It was taking out everything along the way—including large stands of ponderosa pines. Wild life was suffering fiercely, for observers were reporting deer, elk, fox, and coyotes—among the larger animals—escaping the danger as the fire raked across the drought-stricken forest. Authorities would later debate the feasibility of setting that control burn, which obviously was no longer controlled nor just a burn. The fire licked up tall evergreens in seconds, as it gained speed in the fast prevailing winds from the west with gusts over 30 miles per hour. As it grew, the fire created its own weather, and tornadic spirals of fire rose forty feet into the sky.

Mel remembered taking his family from Albuquerque one summer and visiting a local, rimmed valley—the Valles Caldera—in these mountains. The circular rim is nearly 14 miles in diameter and encloses a volcanic dome. His children were thrilled to see the grassy valley encircled by evergreens. Now, from the reports, a part of those encircling stands of trees had been burned out as the fire marched across the forest.

As they worked, a supervisor kept them informed on the progress of the main fire. Finally, they were ready to set the backfire, for the winds had, as least momentarily, shifted to the west. Behind them they had a city with its homes, and even a federal laboratory. Mel learned of the evacuation orders that had been given to residents, and he thought of the beautiful forest surrounding the city. Placed on a plateau with deep escarpments forming canyons within and around, the city was high above the river valley to the south. The people were depending upon these firemen and their work.

The order came for them to clear out to the base camp. Catching a ride, Mel looked over his shoulder as he saw a couple of firemen setting the reverse fire. The westerly winds that had been to their backs continued, and as his truck rounded a bend, he could see the new flames licking toward the forest to the west. Exhausted, he came to the camp, shed his equipment, and found some refreshing water to drink and wash his face. His canteen was nearly empty, and though his truck had a water canister, the larger supply at the base camp was very refreshing. He caught up on news of the fire's progress as it devoured thousands of acres of timber and grassland between the stands of trees. Finally, he managed to make a brief call on his cell phone to his wife back in Albuquerque. "Yes, I am safe," he reassured his concerned spouse.

"I'll be praying for you and the others as you fight this disastrous fire," she assuringly responded as she choked back sobs of fear, which she tried to conceal from the phone. Mel understood, for they had been through this any number of times in the past, but this time it was very big. The several-years drought and the rising temperatures of the summer had created the conditions for this blaze. Fears of dry thunderstorms with lightning had prevailed even before the crew had set the original control fire. Those storms could come up and sprinkle a few drops or only create virga—drops falling but never reaching the ground. Then a lightning bolt would hit the ground, ignite a tree, and the fire is on. And Mel and his wife were thinking about that as they talked of the current conditions.

"The heat's on this time," Mel shared with his wife. "That may sound trite but in places the timber is even drier than what we burn in our fireplace. I'd better get back because I see some excitement on the other side of the camp."

Another truck had come down from the firebreak. The driver jumped out. "They have set the backfire, but reports are that the wind is reversing, and the main fire is coming faster than expected." The nervous men tried to move methodically, but the thought of that fire coming down the mountain haunted them. Suddenly, the silence was broken.

"The fire's on the way," came the announcement. "Clear the camp!" With that, men were grabbing gear and throwing it into trucks as quickly as possible. Within a few minutes the essential items had been cleared from the area, and the trucks were grinding down the rough forest road.

Mel's wife saw the report on her TV. The wind had shifted, and the fire came over the ridge near where her husband had been working. It jumped the firebreak and marched on. A very concerned spouse finally got a quick call from her husband. "Honey, this is Mel. We're safe, but the fire is hot and dangerous. We're being reassigned, as I talk. I must go!" She managed to give him a reassuring "I love you" just before she heard the phone go dead.

Fighting exhaustion, Mel had worked for the entire day and now into the night. No real rest was coming for these very tired fighters. The fire marched down the mountain side and into the outskirts of the city. Airplane drops of fire retardants had been part of the preventive effort even while these men had been working on the firebreak and backfire. Safety conscious supervisors were not going to let these men experience the disaster of several recent fires when fighters were caught in the advancing blaze and died. Now the drops of fire retardant intensified, as other resources from neighboring Colorado and Arizona had been called in. This was a fight for preservation—of a city. Mel's department and others went to the western outskirts of the city and tried preventive hosing down to put out sparks and even bombs of fire coming from the now roaring caldron to their west.

Tall trees had been taken down west of the city several years before this marching maelstrom came to the city's edge. That space finally saved the majority of the homes, but the sparks and fire bombs outnumbered the weary firemen, and some homes did burn to the ground. With the gargantuan efforts of the fire departments from the city and all around—including Mel's Albuquerque crew—the fire was finally being slowed down and even suppressed in places. Fire crews did manage some shifting once the disastrous threat to the city had been averted some hours after Mel had been transferred from the mountain. Exhaustion put him to sleep wherever he was able to drop, but the sleep was restless, for he could still hear all the commotion of trucks and men coming and going.

After three days, the caldron had been conquered, and the small tufts of fire that tried yet to live in the forest and some neighborhoods were snuffed out. Mel, meanwhile, had continued to be in touch with his wife whenever a moment at his camp allowed it. He had survived, but one very weary husband finally after almost a week of extreme hazard returned to Albuquerque with his crew and virtually staggered into their safe home. Mel had been part of many fires, and even working firebreaks, but he would never forget this fire and the fate of their *backfire*.

8

Windows

Setting Arizona: Urban Home. Twenty-First-Century Great Recession

Windows—"Those are the next things that will be changed ***and*** charged," Julie thought, "in this twenty-year-old house on which we just floated a financially death-defying loan." The view from the front bay windows was beautiful, she and Bob both had agreed, when they first looked at it. Across the street from a lake—*in the desert*—the vista laid out a golf course with the lake between them and the eighth fairway.

The view was indeed very lovely, but knowing those windows let out cool air in the summer and let it in during the winter wasn't energy-efficient—and they were "green people," energy-conscious and so wanting to save the environment. They had talked to the realtor who worked a deal with the original owners by reducing the price. "It's too good a deal to pass up"—that's what they both thought. The price all meant less on their mortgage, but still a hefty bill for replacing those "energy-filtering" windows. They had scrimped and saved just to put together a down payment, but there still were those windows. "We're going to pay through the ceiling for our electric bill," Julie lamented, even though she loved the home. They certainly knew what those bills could be, for they had moved in the middle of the summer with record-breaking temperature.

"Oh, well, we found work here and loved the desert when we came looking last December." That's what she used as a rationalization as they sat in the heat of the summer—looking at that view through those windows. As mundane as it might be, those panes needed to be changed and fast. The couple resolved to do it, but where to find the money. They had budgeted to the last penny and would be living from paycheck to paycheck for some time while they recovered from the cost of the move. "But, maybe there's a place for one more payment," she thought. Anyway, that first weekend in their home—still surrounded by boxes and misplaced furniture—they were celebrating—they had jobs and a home. A bottle of wine and a full meal from their kitchen made things "feel like home, *finally,*" she thought.

In the meantime, they were settling into their new jobs—careers, they hoped. Julie had signed a teaching contract at a nearby high school, and Bob was working for a building company, where he would be a site supervisor—very much in step with his prior work as a contractor. However, here there was work for him while back in the Mid-West the bottom had fallen out of the housing market and his contracting business had wilted, just as had the economy. They came home each day to the boxes and plans to arrange the little furniture they had brought from their apartment "back home," for they had both grown up in small Mid-Western towns, met in college, married, and found jobs near their parents' homes. Then Wall Street greed and Federal favoritism for the wealthy wiped out the economy, and this middle-class American couple had to move. "We must make the break, since Bob's work has died," she had explained to friends—not placing blame on anyone.

"You're young, and we understand," her mom had said, but that hadn't stopped the tears as she had turned away from Julie for that final "goodbye." They had been close, for Julie was their only daughter and had married a man "from the area." They all had hopes of settling into a small-town life in the Mid-West, but that was not to be with the economy taking such a dive. They had heard of the Washington conservatives' favor with the Wall Street speculators who had messed up the real-estate world with their creative—and illegal—financial packages. "We're not going to be part of that mess," she had assured her mom. "We just want a new start and a new home!"

"And that we have," Julie tearfully recalled as she sat on a box one day after work and gazed out their bay window to that lake across the street. She was sad for the loss of friends and family nearby but relieved and—yes, happy in a way—for having this new setting. She and Bob were going to make it! Arriving home from work, Bob broke her reverie, as he came into the room, and glancing across the street remarked, "You know, I still can't quite adjust to a lake in the middle of the desert. I know what they told us about recycled water and all that—and I'm understanding—but it still seems inconsistent and out of place. They use the water to irrigate a golf course, I know; but with the drought, how many golf courses can we have here?"

Changing the subject, he continued, "Hey, I did some checking today through some guys at work about these windows, and I think we can work a deal with one of the subcontractors."

"Oh, really? But there's still how we're going to pay for it," Julie added, almost interrupting her husband.

"Yeah, I've been thinking about that, too," Bob was quick to respond. Money had become a "touchy" subject for these two—obviously because of the cost for their transition and the slowly recovering economy. "I think we can work something out through one of the financing banks that service our clientele. Interest rates are 'dirt cheap' at the moment, and I'm really glad we didn't put the cost of new windows into our mortgage. We can work something out with one of those bankers, I'm sure. I'm having lunch with one tomorrow!"

"Okay," Julie said in a rather nervous and acquiescing tone. That assuaged her concern, and they both turned to watch the tree's cast their shadows over the lake. The couple had been advised to buy a north-facing home to avoid the extreme temperatures of an east-west orientation. At this moment, they were enjoying that advice as the setting sun cast its beams from their left over the trees and lake and turned the cumulus clouds above them to bright yellow and orange.

The next morning, Julie had a new-staff meeting to attend at the high school, and Bob went off to his contracting site. At noon, he met at a local restaurant—as he told Julie he would—with a banker. In his own jovial way, Bob put the banker at ease and explained their predicament with those single-pane windows, the need for replacement, and the money to pay for them. "Why, that's a minor matter," the banker reassured Bob with a smile. "Wow, that was easy," Bob thought. "Just contact my office when you the need the money, and my assistant will make the arrangements." The luncheon ended, and both men, exchanging business cards, returned to their work.

That evening, Bob returned home before Julie had finished at the high school and decided to broil some salmon that they had managed to buy between all the moving and work responsibilities. One of the first "garage projects" for Bob had been to get his gas grill reassembled. So he used it this night on their patio. He heard the garage door open and close and Julie's greeting. She dropped her briefcase on the couch and, going outside, declared with a sweet surprise, "My, aren't you the chef!" With that, she went closer and pecked him on the cheek with one of her affectionate kisses.

"Good news! I met with a banker today, and we have the money! The interest is low, and monthly payments will be affordable for our budget." That was sufficient for Julie. "Oh, and a telephone call to one of our window contractors sealed a deal to make the changes. He'll come by over the weekend to look over the house and take some measurements. The replacements will be two-pane, vacuum windows and, I'm sure, will save on our energy bills."

"Great!" With that detail, Julie knew that they had one less financial thing to worry about. "And now what do we have for dinner tonight? What can I do?" With that, they turned to the evening, to their new life, and to no more worry about the *windows*.

9

Coal

Setting Arizona: Navajo Nation, Kayenta. Twenty-First Century

Coal—that's what she was addressing! A federal employee, she was facing the coal lobby. Anyone with a lick of sense realized that industry could not keep belching coal soot and carbon wastes into the air! Hired by an appointee of the President, she was working in the Environmental Protection Agency—EPA, to the Feds—and had been involved in current research on effects of coal on the atmosphere—particularly in the Southwest. A distant relative of the Navajos, Maia—or Mai meaning "Bright Flower" in Navajo—was proud of her Native American heritage and had been in contact with some of the current leaders of the Nation. At the moment, she felt alone—except for her colleagues at the EPA. "Where can I find someone who knows and lives with the situation?" she asked herself almost everyday, it seemed, as she went to her Washington job and thought of having to face that coal lobby, *again*.

Her employer, the EPA, was no favorite of Washington Congressional conservatives. Their candidates had threatened de-funding the Agency in campaigns for the Presidency. One Presidential candidate even had maintained in a televised debate that he was going to eliminate several Federal agencies, but could not remember one that he had included in the list—the EPA. The coal lobby had been her nemesis as soon as she started checking into the energy question. In the office, she did research on the coal companies. Off the job, she had been checking into some of their plants—particularly in the Southwest—asking questions of some environmental contacts. A couple of telephone calls to some friends who had lived in the Southwest brought some details about which she was shocked to hear, but perhaps not too surprised when thinking about what the lobby and industry could do. "I can't say that I didn't expect it," she said to herself as she heard them relate some of the effects of the smoke on the Native American children. Additionally they talked of disposal problems.

Well, the coal lobby and empathetic conservatives wanted to preserve big business and not spend any more money than necessary to turn out the product—she had quickly learned that in her new job.

In the coal mines of the Eastern United States, some coal-mine owners had put off government inspectors' demands for improved safety conditions. In a few instances, with coal strikes and cave-ins, Maia remembered, these owners had just shut the mines down. In the case of coal for power plants out West, she knew that more scrubbers would be necessary to eliminate harmful emissions from the coal plants generating electricity on the Navajo Nation. Rumors had it that filters currently used, when filled, were turning up dumped along roadsides on the Nation's land or in old, abandoned houses.

Through this federal job, Maia had learned much about all that and more when it came to what the coal industry had done in the Southwest. She knew about the history of the tribal holdings and what had developed over the years. The Hopi Tribe together with the Navajo Nation, she had found out, had holdings of coal on Black Mesa, a high plateau of 5400 square miles in Northeastern Arizona. The Hopis had actually been extracting coal from the mesa top since about 1300 A.D. for making pottery and heating their dwellings. Around 1900 surveys resulted in some small mines being opened that provided locals and schools with coal. Trying to assist Native Americans, the federal government then passed in 1950 the Navajo-Hopi Rehabilitation Act, which was to assist the impoverished tribes with development of their resources. From that Act had come the development of coal mining studies and eventually of power plants. "Of course," Maia thought, "this provided energy companies the opportunity to be away from metropolitan areas with their polluting smokestacks."

She knew that the coal industry had welcomed the opportunity to establish electrical generating plants in these remote areas of the Southwest to "help" the Native Americans and, above all, to "help" the coal industry's pocketbooks. They mined coal from Black Mesa and other sources on Navajo lands and transported the coal to plant sites. On Black Mesa, the mine near Kayenta was producing bituminous, or black coal. A relatively soft coal, the bituminous was still of better quality than lignite coal. However, Maia had known from her college days that anthracite, the hardest coal, was the least polluting of the three kinds. "This soft stuff is particularly polluting and requires filters," her professor had told her class. So when she took the job with the EPA, she learned more about the coal company on Navajo lands and its current operation. The Black Mesa mine had started operation back in the 1970's and was supplying an electric-generating plant with the coal as fuel to power its steam-generators that produced the electricity.

As an environmentalist, Maia understood the United States challenge for energy. The Southwest was growing rapidly, and hydro-electric power had been maximized. Another dam on the Colorado River just above the Grand Canyon had been stopped years ago, and something else had to be done. This EPA employee realized that a transition fuel would be necessary to supplement Western hydro-electric plants and to move from carbon-based energy to so-called clean fuels such as solar—including photovoltaics, thermal collectors, wind, and biofuels. In renewables, large production was being studied and slowly but surely developed through the Department of Energy, but right now something had to be done about these coal emissions.

Once more to obtain first-hand information, Maia planned a trip to the Southwest. Landing in Phoenix, she rented a car and headed north to Navajo land. In prior conversations with her friends and a couple of like-minded colleagues from Arizona, she had heard of a State legislator from northern Arizona and her concern about the energy situation. This Washington EPA "bureaucrat," her friends had teasingly called her, was anxious to meet the Arizona legislator and to learn more of her background. Maia now had been in her position for a couple of years and, not completely realizing it, she was soon to meet a like-minded woman. Now they had a meeting scheduled in Flagstaff at a hotel retreat center.

While Maia had acquired her education, an internship at the EPA, and after graduation this position with the Agency, a native American girl—a Navajo—Johona Hayou had also completed her education and pursued her political career. One of Maia's Arizona friends, knowing both women and their empathy for environmental causes, felt that they might build a friendship that could bring them to a mutual destiny in meeting and taking up their cause. That was the person Maia was to meet in Flagstaff. Her trip from the Valley took her from an urban setting, on an Interstate Highway through mountains, past saguaro cactus that disappeared around three-thousand feet in elevation, and across rolling countryside all the while climbing to the Mogollon Rim, a high plateau in Northern Arizona. Passing through a broad river valley, she noticed a sign to two National Monuments—attesting to an ancient Native American culture about which she remembered reading. Climbing to the high plateau in the distance, she periodically caught a glimpse of the San Francisco Peaks that towered over the countryside. Junipers started to appear on the knolls along the route; and finally reaching the rim, she thrilled at where the ponderosa pines took over.

"What a beautiful trip," Maia thought as she spotted a highway sign indicating two miles to Flagstaff. "This is part of the perks for this job," she mused, "but there is the real reason for this trip—to meet this Navajo legislator and environmentalist and to learn what she can tell me." Leaving her route north and moving to another Interstate Highway east-bound, this traveler saw her exit and easily found the hotel. Checking in, she found a message. Maia settled into her room and telephoned her contact. "Let's meet tonight in the lobby and grab a bite here at the hotel," Maia suggested. A brilliant sunset greeted Maia as she left her room and headed for the lobby that evening. She entered the high-vaulted lobby and, turning, spotted a handsome, well-dressed woman coming toward her. It was her contact. "You are Maia?" the woman asked in a rather deep and commanding voice.

"Why, yes, and you must be Ms. Hayou."

"Yes, I am Johona Hayou—Jo, to my friends—and am so happy to meet you." Brief exchanges followed and motioning to two conveniently placed leather chairs facing a broad window, Jo followed Maia to the chairs. Comfortably seated and appreciating the last yellow-painted clouds that the sunset was leaving, the two women started almost simultaneously in conversation. They both felt so confident and welcomed the thought that this meeting could lead to a professional relationship with, perhaps even, a friendship.

With the introductions completed, Maia quietly suggested, "So tell me about yourself." With a smile and her usual confidence, Jo responded, "Well, as you may know, my home has been on the reservation. Our nation has been here for a long time; and when I was born, my mother chose Johona as my given name. It means 'Sunny', and for the sake of our white friends, when I entered school, they gave me the name 'Johanna', an anglicized version—at least in their minds. I soon shortened it to Jo, and that's what it's been ever since."

"I'm proud of our family name 'Hayou', for it means 'Durable One'. I have tried to be happy and certainly loyal all my life, for my mother has been the solid rock for our family." She went on to explain details about life on her Nation's land. She had gone through the kinaalda ceremony at thirteen, when she had become a tribally recognized woman. "My mom always told me that from infancy, I had been a laughing, happy child, and so my name is Johona, 'sunny'. Our family name, passed down in our matrilineal society, has been Hayou, 'Durable One', for generations. I repeat this because it gave me the drive to become what I am today."

Recognizing an empathetic and understanding woman in Maia, Jo went into more detail. Appreciating the meaning of her name, she had taken it to heart as she went through school. She boarded the bus everyday and took the trek from the parcel of land that her mother and the ones before her had worked. The women raised the children, tended the land, prepared the meals, and sought income from their artistic skills. Jo's mom had an artistic bent in several areas. She wove woolen rugs, made silver jewelry, and even had tried her hand at "throwing pots"—describing the method of spinning clay into beautifully round forms and, when dry, adding Navajo artistic paintings. Jo explained that she had graduated from high school and went on to the university in Flagstaff and there completed her course of study.

Maia continued to be intrigued with the story and urged Jo, "Please go on!"

"Concerned about my people's needs and realizing that I should find some way to help, I decided to return to my ancestral land and roots. However," she explained to Maia, "I was going to find a way to make a difference. That came through politics, for I had been involved in student government while on the university campus. I chose to run for tribal council and then later for the State legislature." Rolling into the deeper background, Jo continued to tell of her convictions and concerns. Speaking with tribal members about the need to control their own destiny, she found empathetic ears and was elected to the Council. However, she knew that was not enough for her cause, and she chose to take on a legislative position that had just been vacated. Attending tribal gatherings, consulting Council members, and speaking with every Navajo that was willing to listen, she managed to garner sufficient votes to be elected to the State Legislature. Among her major heartaches were those coal-burning electrical plants on her homeland. "They must be cleaned up, I told my mother on more than one occasion," she explained to Maia and then, realizing that she had been talking for over twenty minutes, apologetically interjected, "Please forgive me for getting 'launched', but our early contact made me feel that you would understand."

"Oh, yes, and please continue," Maia responded as both women glanced outside where spotlights on shrubs and trees had replaced the setting sun. "I know you're telling me what I need to know to continue on my quest. So please go on!"

"Well, one day, I remember seeing my mother standing on a very dry barren knoll. I watched the coal soot and smoke blowing over her as she stood there trying to see what could have been a beautiful sunset. That image stuck in my mind and became the determining factor for my mission to be elected to the tribal Council and then eventually to the State Legislature. Something had to be done to change all that. And here I am."

"And I fully understand, Jo," Maia reassuringly responded to this Navajo legislator's account of her life and determination. "Beautiful, and I think it's time for dinner," as she glanced toward the hotel restaurant. "We can continue our conversation over a meal," the Washington visitor suggested. With that, they went to the maitre d' and were ushered to a table.

Without hesitation, Jo then asked, "And how is it that you came to this place in your life?" With that introduction, Maia went on briefly to explain her educational path, vocational career, and her own intent. Explaining the purpose of her visit in more detail, Maia found an affinity with her newly met acquaintance. This EPA employee was impressed with the fact that she was meeting on land adjacent to a Native American nation that covered parts of Arizona,

New Mexico, and Utah with over 300,000 residents. They continued their discussion over dinner and grew in understanding their mutual concerns with the environment and women's causes. "This matrilineal society underlines the need in our society today for growth in careers for women," Maia thought, as in a moment of quiet she finished her meal. Concluding their dinner, they agreed to meet early the next morning after breakfast for a day of touring and consultation.

Later that night Maia mulled over what she had been told. She knew this was the first of many contacts in the future. "These women work the land and raise the kids, thereby preserving their culture. They make pots, weave rugs—as they have done for centuries. They own the land and livestock and then train their daughters to know how to manage it all. That's what she's told me." Neither woman slept very well that night, for they both knew that they were onto something.

That next day was filled with stops in the Navajo Nation, as Jo took her Washington visitor on the brief tour. Maia had opportunity to see the power generating plant in the distance and on Highway 160 to view the coal operation from outside the site. Seeing the details and touring these facilities would be left for another time, but now she had to know more about Jo and what was happening on the reservation.

Jo was proud of the opportunity to introduce her mother to her newly found friend from Washington.

The three women were quickly drawn to one another, and Maia learned and saw the rudimentary existence that Jo's mother had. The government had provided housing, but the stark, barren land around the dwelling was sealed in Maia's mind. Now she appreciated the image of the mother on the knoll—under the smoke. The two younger women left with promises of returning again soon to the mother. On the way back to Flagstaff, Jo took a detour to show Maia the stands of pottery at a spot along the Little Colorado and drove on into the National Park to see the Grand Canyon—a "touristy thing," Maia thought. It had been a very long, but profitable and enlightening day, and they returned late that evening to the hotel in Flagstaff.

"Let's meet tomorrow here at the hotel," Maia suggested. "I can make arrangements for a conference room, and we can bring our laptops to do some strategizing."

"That's fine with me. See you then," Jo responded as Maia exited Jo's SUV, left the portico, and headed for the main lobby and her room.

That next morning started the planning and strategies that would take both women back to their jobs—EPA in Washington and the State House in Phoenix. But this was only the beginning. They discussed Maia's next trip to Arizona and an official visit to the power plant and coal operation. Both knew what they would be facing—corporations, coal lobbies, conservative legislators, and rich, self-seeking power brokers whose main interest was their shareholders. But that was part of their environmental push. Maia realized that from the EPA could ultimately come a proposal for a directive from the President to require the coal-burning powerplants to provide more scrubbers within, hopefully, the next eight years. The two women had not seen any dumped filters on their tour, but in the future they would be able to talk to Jo's friends and family who knew about what was going on.

After their morning session in Flagstaff, Maia bid farewell to her new friend. Over the brief visit, they had opportunity, between all the professional exchange, to speak of their personal hopes and dreams and to have some relaxing time together. More of that would come, but now they had their tasks to work on—changes for the better. The legislature could work to have the state bureaucracy insist on cleaner air—not just for the urban areas, but also for the rural spaces of the state. The EPA would have more detail on the environmental issues with coal in the Southwest. Both women realized the camaraderie that was theirs now and would be through contact with visits and communiques in the future.

As Maia retraced her drive back down the Interstate later that afternoon to make her flight in Phoenix—warmed by the memory of her newly found friend—she had this overwhelming image of a matronly Navajo woman standing on a knoll in the downdraft of a coal-burning power plant. "This scene must be cleaned up," she thought. "If coal must be used as a transitional fuel into the future, why can't they use improved scrubbers to change this picture now?" With these images and many facts swirling around in her head, Maia returned to her job later in the week—working toward a future when renewable energy would be more prevalent and, at a remote power plant in Navajo country, utilities would no longer be burning *coal.*

10

Irrigation

Setting California: El Centro. Twenty-First Century

Irrigation—that's what he'd been doing all his working life—moving irrigation pipes. He'd been planting the same vegetables every year—lettuce, carrots, cabbage, and other truck-farm crops. Along the way, he'd managed to secure a plot of land with the water rights. Though struggling most of the time to make ends meet, he was able to marry and take care of his family. Now the water company had informed him that the water was being cut off. As so-called "truck farmers," these growers were part of a partnership that carried its own brand name. Once desert, the area was transformed with the coming of the big canal and water to the parched land. Now, everywhere he looked he could see land that had been leveled and banked with irrigation ditches routed to the fields. Again, however, they were returning to the parched desert from which the acreage had come. A number of his neighbors had given up because the cost of fertilizer and seed and labor had eaten them up, and they just drove away. Now he was faced with this news that there would be no water at all. The drought had taken its toll. What was he to do? He had a wife and three children—how would they survive?

They had received some rains in the last monsoon, but that was not nearly enough to refill any reservoirs. Anyway, they were getting water from the Colorado River. However, a treaty with Mexico, an agreement between the Southwestern States on prescribed acre feet, and urban needs had all intervened. The last of the trickle that they saw in their main irrigation ditch was sufficient indication to note that the remaining water had been allocated. Besides, their annual rainfall was only six inches—a desert climate—and now that wasn't coming. Plenty of clear skies for sunshine and mild winter temperatures—that all contributed to winter crops that supplied grocery stores "back East," especially. Recent summers, however, had been very hot, and the last two were scorching with record heat. The monsoon shower had been only a sizzle on the dry ground.

The evening that notice came, he had walked from their unpainted home down the dry, rutted road to the mailbox—all the while glancing over the parched land and thinking about his shriveling crops. He had collected the mail—the usual farm magazines, some advertisements, a monthly letter from church, a bill, and the one that caught his eye. The official nature of the letter couldn't be lost in the pile of routine junk mail. He had to open it before he got to the house, and the shock came as he read the first lines—NO WATER—that's what jumped out after reading the first two sentences. Quickly perusing the page, he shockingly learned what he knew would come—the irrigation company could not supply any more water.

He just couldn't help himself—he ran to the house, opened the screen door, let it go with a bang, and called, "Ginny!" The urgency and excited tone brought his wife quickly from the kitchen where she was preparing the evening meal.

"What is it?" she questioned with an anxious frown, for she knew that her husband Guthrie was ordinarily the cool one. He took things with a certain casual offhandedness usually. The shocked look on his face and the waving of a letter in his right hand presented an image of panic to Ginny.

"What in the world is wrong?" she asked with an earnestness that she rarely showed to her already worrying husband. He had enough on his mind with trying to grow crops, pay the bills, pay hired help, and wonder about a water supply. And now he comes almost shouting into the house!

"We got a letter from the irrigation company! It's official. They can't provide any more water," he moaned as his eyes filled with tears. It was the first time that Ginny had seen her husband cry. The shock was just too overwhelming. "Those danged city slickers are taking our water for their lawns and fountains. They don't think about their food supply. Just the water—that's all they care about." The truth be told, the urban areas had used water to irrigate the trees and shrubs along their freeways, and the private consumers had watered lawns until city restrictions forbade lawn-watering more than once per week. "Anyway, that's there, and this is here," Guthrie thought as he began to compose himself.

"What are we going to do?" Ginny blurted out. They studied the letter and began to work through their anxiety. The kids had just gotten off the bus and were coming into the house when their parents had reached a level of calm—not resoluteness to the situation. It wasn't as if they didn't know that something would have to happen because of the drought, but this really brought it home to the couple and then to the kids, as well. The dinner that night had a pall over it; their meal prayer tonight was filled with pleas for guidance and resolution. After the dish washing, they all sat down in the livingroom, and they began to reminisce—the good and the bad—a trip to Disneyland two years ago, sharing all the work around their home, stretching their budget for just what they needed, loving school, and just treasuring being a family. Of the three kids, two were in high school, and they were more anxious than their younger sibling, for they knew this might mean leaving friends and sports at school before graduation. They had seen it happen to other classmates, whose parents had to leave. That night a daughter and two sons dragged themselves to bed—all with anxiousness on their minds and nightmares in their sleep.

After an almost sleepless night, the parents did get a little rest and, at least, could think more clearly the next morning. They started to discuss what they might do and to weigh their options. Very little had been put aside for the future—all the spare cash had gone into the farm and what absolute necessities the kids needed for school. One thought was in the back of both parents' minds—what about that offer from Colorado that they had been given over the last years. "Hey, what about checking on that invitation from my parents?" Ginny offered.

"You know I was just thinking the same thing," Guthrie responded. "Your parents have always wanted us to come back to your home. They are getting up in years, and I know it must be getting very hard for Dad to work all that acreage."

They discussed the matter in more detail and agreed to contact Ginny's parents on the eastern slope of Colorado—the Front Range in the Arkansas River Valley. Those had been Ginny's roots before they moved to Southern California—SoCal, as it's known there. Certainly Ginny's parents were aging, and being an only child, she was concerned about them.

"Yes, perhaps this is the time for us to go back home," she slowly resolved. The farm there had more acreage, but the winters could be hard. Stories from her past filled the couple with worry, for they had sought a drier and warmer climate to escape the harsh winters. "Fine thing that decision turned out to be," she thought as she mulled over all the preparation and rigmarole they would face if they moved. Certainly the couple had discussed, before their marriage, their own worries about the troubles of their parents who had come from the Mid-West to the Front Range of the Rockies and the Arkansas River Valley. That parentage had found better land and more water than where they had been. This California family knew that the Rockies did have snow on the Front Range to supply the Arkansas River. That river didn't have the urban demand on it near its source in the Rockies that the Colorado River itself had as it flowed through the arid Southwest into reservoirs and back into its canyons and finally to the Gulf of California.

Well, Ginny and Guthrie knew that the drought was hurting Colorado but not as severely as in Southern California. Here in California, they lived near a town that actually was below sea level with the summer temperatures sitting over 110 degrees for days on end. "Let's see what we can do," Guthrie finally offered after their long conversation that morning. The kids had already gone back to school after their early breakfast—had gone back with great reluctance because they were worried about their future and what they could do to help their parents.

Ginny was the designated one to check with her parents for what could be good and bad news for her parents—good to have the help but bad to know that they were losing their farm. That afternoon she got up the courage and called her parents while Guthrie want out with heavy heart to his usual work—chores now—around the farm.

"Hi, mom, how would you like some company?" Ginny started as she heard the familiar voice on the other end of the line.

"Well, sure! You have a standing invitation to come, but aren't you in the middle of your growing season?" came her mom's response, while she looked out the window over a snow-covered farmyard in eastern Colorado.

"Mom, I have some good and bad news. The bad news is that we are going to lose our farm. The drought has gotten us. The good news is that we would like to come back to Colorado and help you and dad on the farm. Is the invitation still open?"

"My goodness, yes!" came the blurted response without hesitation. "What brought this on?" Ginny then went on to explain the letter from the irrigation company.

"We've known this could come, if there wasn't more snow and rain in the Colorado River Basin. It's here! They told us we wouldn't have any more water. Our well can't begin to irrigate our fields—it's barely enough for the house and our animals. So, we'll write you about our plans, as soon as things are more solid."

"Wow! What a surprise this will be to your father when I tell him. Take care of yourselves, and we look forward to hearing how you're going to do it."

So the plans were made, the letters were written back home, the notice was given to the irrigation company, and the farm workers were told what they had suspected for a long time. Guthrie had always wanted to be fair to anyone whom he hired and pay them a living wage—unlike some of the politicians from a couple of neighboring states who hated a minimum wage. "These are people with families just like us," Guthrie would tell a couple of his neighbors in the rare moments when they were together in the middle of the summer with opportunity to visit.

The couple knew that the sale of their land was nigh unto impossible because of all the prior desertions of land by their neighbors. On the way through their town, travelers could see the parched land with berms of earth on all four sides of each flat field and the dry ditches that once carried water to them. Guthrie checked with a realtor and notified the bank. He felt deep in his bones that the land would not sell and would eventually revert to the bank. "I can't just walk away from this place, for Ginny and I have put a lot of blood, sweat, and tears into making this a home for us and the kids," he thought, as again the tears trickled momentarily when he stood on the edge of a parched field of carrots that started to grow but never quite made it.

After all the notices and goodbyes to the neighbors, the day finally came when Guthrie went with his oldest son for the two trucks that would carry their household goods to Colorado. The kids had managed, though in mid-term, to set up transitions to a school in Colorado. Grandma and Grandpa back on the farm in the Arkansas River Valley were anxious to see their grandkids, but those kids—though loving their grandparents—were filled with real trepidation on what they would find on the other end of this long journey back to their parents' past. While the other animals all had been sold for a pittance, the kids just had to take their dog, the "other member of their family," as the faithful Shepherd had been designated by everyone in the family. The three remaining farm workers had insisted on coming to help with the loading. Ginny managed to put together a lunch for everyone. Packing had preceded all of this, and by mid-afternoon on the day of their departure everything was loaded onto the two trucks. What a tough parting this was! Grown men hugging one another with tears in their eyes as the three faithful farm workers bade farewell to their former employer and hesitantly but firmly also embraced Ginny, who approached them with outstretched arms to say goodbye.

The drought had come and taken them—another family that fell victim to the climate change which inevitably was placing a pall over SoCal and the entire Southwest. Guthrie and his son each climbed into one of the two rented trucks, while Ginny went to their weary and worn pickup and their daughter to the sedan with her younger brother. With dad in the lead, this caravan of vehicles with waving to the onlookers went down the lane for the last time, past the "for-sale" sign, onto the cracked but paved road that led to the Interstate that would take them to Colorado and hopefully a better life. The aluminum pipe that had brought water from the ditches into the fields glistened in the afternoon sun, and Guthrie thought, "Here's one more place where there's not going to be any more *irrigation.*"

11

Climate Change

Setting *Texas: Small Town, South of Guadalupe Mountains. Twenty-First Century*

"*Climate Change* is the next part of our study. Read the assigned pages in the text. Each of you will be picking a topic for research on climate change and its effect on the earth." Elizabeth—"Liz" to her family back home and "Ms. Scherer" to her students—was explaining the next unit in their general course on science. Her high school, in this small town in West Texas, served not only the community but the outlying ranch folk within this part of the county. South of the Guadalupe Mountains, the area had once been more prosperous as ranching country with the town serving as a source of groceries, dry goods, and fuel—especially on Saturdays. Now very dry and a victim of drought, the countryside appeared barren and "wanting rain" to any visitor who drove through or stopped for fuel at one of the few remaining service stations. As she concluded her last class of the day, Liz looked out a very dusty window to the dry landscape, and her momentary thoughts and doubts all returned.

* * * *

"Whatever made me come here? Can I really stay?" Liz had asked herself several times soon after arriving. Visiting was one thing, but settling into a small-town environment was still another thing and certainly very much in contrast to her roots. However, intent on serving in a school district with less means and appeal than suburban areas, she had sent out resumes to several Southwestern state school boards while checking the usual notices published in the professional sources available to job-seeking graduates. She had been, of course, applying for a teaching position in secondary science. Several responses had come back with offers that reflected the limited financial resources that small school systems typically have available.

One offer stood out, however, for its appeal and the real need of the school. The principal's accompanying letter explained, "We have had a volunteer staff member teaching science with little background in the field. The limited salary has caused new graduates to move on within a year or two, and our high school students really need continuity. Our small faculty's goal has been to improve our secondary program so that we could send more students to the State's fine universities." The principal also had written of their program of kindergarten through high school being housed in one large facility. "As students progress through our system, the classes become smaller because kids have little incentive to finish even high school," she lamented in the cover letter. "They move on to service jobs or back to the ranch from which they came."

Liz remembered being moved by the principal's appeal and on her invitation had decided to visit the small town. Almost feeling guilty of taking travel expenses from the school, Liz still said to herself, "I really must see this before I jump in." She had flown into El Paso; and renting a car, she had driven the miles east to what was now her town. On that visit, she remembered how warm the principal and staff had been; she was moved by that and the appeal of some caring parents. So, she had decided that this could be a mission for her—to provide some meaningful and valuable education to their students. "But how am I going to fulfill myself and grow?" she had asked herself. The answer had come when she noted that the school is in proximity to the Guadalupe Mountains, and that fulfillment could be explained in her educational background.

* * * *

"Ms. Scherer?" She heard a familiar voice bring her from the reverie. "We had spoken about getting together today, but I have to go with my mom to the doctor. Could we do it tomorrow after class?" Susan asked. A gifted student in Liz' science class, Susan had been working on an extra-credit project and was particularly interested in her teacher's studies in archeology and geology.

"That will be fine. I completely understand and will look forward to seeing you tomorrow afternoon." As she responded, Liz was reminded of Susan's explanation of how her parents reacted when they learned of her science projects. This teacher remembered her student explaining how the father particularly had responded to all this "science stuff" and what he considered a denial of their religious roots.

Her mother was more moderate, but the father had become very vociferous. They had come to the first Parents' Night, and Susan's father had sought "Ms. Scherer" out of the crowd. Liz recalled the father's words, "I don't want you teaching my daughter anything about evolution and that stuff. Our preacher says the world is six thousand years old, and that's good enough for me and my kid." As she recalled the incident, Liz also remembered the visit after class the next day and Susan's apology for her father and her expressed and continued interest in Liz' science class.

That evening Susan drew off her bookshelf what had really been her personal interest—this science teacher had studied the Guadalupe Mountains, their geological history, and the ruins of some native Americans who had lived in the vicinity. She was planning a trip on a Saturday to do some exploring and mentally wondered if she could invite Susan to go with her. "After all, I could share some of my education and background with her."

Liz remembered how in college a real interest in archeology and geology had developed. On her time off—vacations—from school she had gone on remote digs in the Southwest—assisting university professors specializing in pueblo history. They sponsored archeological digs and invited other educators to spend time assisting at the sites with digging, collecting, and evaluating artifacts under the professor's guidance. That took her to old pueblo ruins with spade and trowel and a lot of patience. The summers were getting too hot to volunteer her time in late June or July (when she was on her summer break), for these digs took them at times into the recesses of Southwestern government lands or even, when permitted, onto Indian nation's country. The drought had brought such broiling sun and temperatures that she only wanted to spend some time over the holidays of her school year to volunteer at the digs—when one was sponsored. And that was what had introduced her to West Texas, New Mexico, and Arizona—thus, the decision to accept this teaching offer to a small town south of the Guadalupe Mountains.

As she perused the materials from her bookshelf, she came across some items from her home. She had come from the state of Washington—Seattle, actually. She had appreciated the drier climate of the Southwest—before the drought and heat waves set in—and really wanted to become a science teacher presenting the history of the discovery of the elements. "Classes in physics and geology in a high school would be great," she had thought at the time, but she had to modify that when she chose a very small school. Earth science would be a very basic course to teach. "I can settle for that," she had thought at the time. "Perhaps there will be a gifted student to whom I can introduce geology and physics." And Susan came to fit that bill, for the teenager had an inquisitive mind and wanted to know more.

As Liz sat at her desk and looked over those bookcase materials, she was mulling over in her mind other courses. At the moment she was teaching this course on Earth Science, but she was working on a course on Geoscience that would integrate several scientific disciplines. She came across the pad and the notes to herself about the new course for the next semester, and some correspondence fell from the stack of papers.

Here was the material that she had collected on her new professional responsibility beyond her teaching—even this early in what would be a long tenure as a science teacher in Texas. Before coming to her new position, she had been in correspondence with a couple of friends who had come to the state as secondary teachers. Knowing of Liz' academic record, one of these friends—without consulting—had appointed her to a curriculum committee that this friend chaired. Liz was to serve on a group for science curriculum, advisory to a state legislative committee. Some legislators had wanted to change the textbooks for the state. Having Liz' academic records and all the awards and recommendations, Ms. Jonas, her school principal, had signed the necessary letters of recommendation and permission when she had been consulted about the new science teacher. Knowing of the ultra-conservative convictions of these state legislators and reading about their declarations against science—including global warming and evolution—the principal had suspected the legislators' motivation and was pleased to have her new science teacher appointed to the committee. After all, the appointment would bring more recognition to their small school district. Besides, Ms. Jonas also had attended another school district's Parents Night and had heard their state representative's tirade against "climate change" and "evolution."

At this moment as Liz returned to the papers and mulled in her mind, "This is going to be a lot of work, but I am up to the task. After all, it fits into all that I have studied and want to do for this school district." She stacked all the papers neatly back on the shelf and returned the books to their place. Reluctantly leaving her desk, she went to the kitchen for some milk and turned to the bedroom for a restful night of sleep.

* * * *

The next day went as usual, and Susan had her Earth Science class in that last period of the afternoon. Susan remained after class and raised some questions about the research project on which she was working. Then, Liz had an opportunity to ask her, "Would you be interested in going with me to the Guadalupe Mountains next Saturday—just for the day? I am going to take my camera and a digging tool or two along to do some exploring and sampling of the rocks in the area. I'd love to have the companionship, and we could become better acquainted." Liz had plans to introduce some ideas about university study to this senior. "Of course, you should ask your parents if it's okay."

"I'd love to do that!" was the quick response from Susan. With that, the two women parted, and Susan quickly left the room to catch her bus to go home. At home, she hurried to find her mother and for permission to go on Saturday. "Oh, mom, I would love to go to the mountains and learn more from Ms. Scherer." The latter reference made her mom nervous—knowing how her father felt about "that new science teacher" and "what garbage she was teaching our kids." This town had Anglo roots going back prior to the Civil War, and Susan's dad was part of that heritage. Her mom had come to the town as a visitor, had decided to stay, and had worked in a local grocery store before she met Susan's dad. The mom realized that the locals were proud of their heritage and suspicious of any outsiders who propounded anything contrary to what they had heard for generations from their parents and the church. Well, her mom, being the more accepting of the parents, agreed and then thought how she would explain to her husband the permission when he got home. The one advantage the mom had was that her husband, for all his gruffness outwardly, loved "both his girls" and left the "raising of that teenager" basically to her mother.

So the next day Susan was able privately to explain to Ms. Scherer that she had permission "to go to the mountains on Saturday." Mom had an opportunity to explain the trip to dad and paint it in the need for their daughter "to have a friend who is a teacher." Well, dad was not too happy about that "teaching stuff" but accepted the idea of a new friendship for his daughter. After all, he wanted his offspring to be a college graduate someday—which he never got to do, being poor and without the money to go. A typical conservative, her father had hated anything that contradicted what he had learned as a child in his religious roots, but somehow it was different now—especially as a father. Maybe this would be a way for his daughter to "amount to something."

Plans were made for Liz to pick up Susan at her home on the edge of town. Liz had packed a lunch and some snacks for the day, and they were off. As their car traversed the patched highway, the two young women were politely chatting about their night and the coming day's trip, but they soon started chattering away about their lives and plans and dreams. Periodically conversation turned to the torrid landscape now without cattle, for the ranchers without water and funds for hay had been forced to sell all the livestock at a great loss. The two travelers saw the parched and cracked land without grass. Within a short time, they arrived at the junction of the highway that took them to the base of the mountains. Theirs was a splendid day of getting acquainted, taking pictures, exploring rocks, and talking of college—including geology and physics as future study for Susan. And they would take a number of other trips to those mountains before Susan went to college.

There would be plenty of other time for that committee back in Austin. The Christmas recess would provide time for Liz to meet with that state committee and challenge those legislators with science and maintaining valid textbooks. Those "guys" would argue with her about climate change being only "just another drought" and not global warming, but Liz and her academic colleagues would prevail. After all, she even had a father back in her small town that might consider some new ideas that his daughter was bringing home about their world and this global warming.

Today, however, was theirs together—the beginning of a teacher-student relationship that would go many years into the future. And Liz had one more person who was concerned about what they could do about global warming and *climate change.*

12

Below the Dam

<u>Setting</u> *Colorado: High Plains, River Country. Twenty-First Century*

Below the dam—that's where they had camped the last time they were here. That time she was quick to tell her husband that she was never going to camp below a dam because she had heard too many stories about flooding and loss of life. In fact, they had seen a campground on the West Coast where campers had been wiped out by a stream. "That's not going to happen to us," she had insisted as they had driven across the country from their apartment in Chicago.

My, how the years had flown by. The first time they had covered a very long leg of their trip, had found the campground in their camping book, and finally had come in very late at night to the site. They had wanted to see the lake by the dam and enjoy another beautiful spot in their trip to the Rockies.

Filled with reminiscing, this time they had come back to the same campground at which they had been many years before. That first time they had been afraid of water levels in any lake in the West because it had been an unusually wet spring and early summer. Runoff in the West over hard, dry ground had been heavy, and stories had reached them in their urban apartment—*that time*! This time the stories were different! What would they find at the lake?

A decade of drought had occurred, and no relief was in sight. In their urban setting, they still had running water—plenty of it, especially with Lake Michigan just a few miles away. The first time, years ago, they were in a tent; and it was easy to set up camp.

"Remember how we got here so late that first time and didn't really know where we were?" she reminisced as they were driving off the highway and into the entrance to the camping area by the dam. "We pitched the tent as fast as we could—by that fourth night we were 'experts'," she noted, as she looked into the rear-view mirror at the small trailer in tow this time.

The first time—fifty miles away from the campground—they had found a fast-food hamburger joint that looked busy and was near the filling station where they had stopped for gas. An hour later—weary and hot because the car had no air-conditioning—they had pulled into the campground. "What a day that was," he remarked as he recalled with her the circumstances of their first visit. "We even survived without many mosquitoes that night!" he added. They had slept solidly and woke up the next morning to what sounded like rain; yet they could see shadows of leaves from the early morning sun. They had camped under a tree; and curious about the sound of drops, she stuck her head out of the tent only to discover that they were under an aspen tree—a *quaking* aspen, to be more exact. The leaves had rustled in an early morning breeze and had given the impression of rain on a surface, a tent, in this case.

"Remember how we laughed," she recalled, "when we woke up that morning—thinking that we were in for a day of rain after hearing those 'drops'? It was just an aspen tree! How much we learned on that trip and since then!" The site was only a day's drive from the Rockies and at an altitude where the aspen could grow. "And then I saw the dam—we were *BELOW the dam*, and I had slept there not knowing!"

"That proved to you that we could be safe below a dam. But, you know, you were right, for how many stories hadn't we heard of people who had innocently camped by a river only to find out that water had been released from a dam upstream or had come from a sudden downpour and wiped out the campground?" he sympathetically responded.

They didn't plan it that way, but this time, too, they were tired and late in arriving at the campground. So, they pulled into the grounds and found a spot—even though they nostalgically had wanted to find their original site. "What's the water level on the other side of the dam?" he thought as he parked the rig in a pull-through. "That's gonna have to wait until tomorrow! As usual, we've got some things to get done."

Again, as last time, they were dry camping—no hookups. Obviously with a tent, one didn't worry about such things, but this time they had tried each night to have water and electric, at least, for their trailer. "We're on the battery tonight," she commented, as she, too, left the van and headed for the rear and the trailer. Not worried about the dam tonight and filled with reminiscing about that first trip, she moved into the usual routine of setting up for the night. His work quickly done—securing the jacks under the trailer and not unhitching for this one-night stop—he joined his wife inside the trailer for a quick supper, shower, and into bed. They had shared the driving for the day and were weary—this time from the added years since they were here the last time.

On a last-minute whim, he had thrown some fishing gear into their van back in Illinois. "Maybe I'll just take the trail across the campground to the dam in the morning and try to catch a couple of fish," he thought as he comfortably lay in their trailer. Well, their night went fast—and they woke up to no quaking aspen this morning. So dark the night before, they hadn't really noticed any of the vegetation. But this morning, when he opened their trailer door, all he could say was, "Wow! Things have really changed!"

Curious at his sudden exclamation, she stuck her head over his shoulder and could only say, "My word! Look at that!" Before them was a *very* dry scene. They had noticed that the ground was rather sandy and dusty when they set up their trailer the night before, but gave it no real thought. Now it hit them! "Lack of memory didn't do this to us!" he said. "The drought has really set in here."

Now shrubs were only sticks set around some waist-high boulders that had marked boundaries for each of the campsites. Trees barely had leaves on them. Certainly it was getting into late summer, but that would only have meant turning leaves—not, "no leaves at all," as she exclaimed and they both took in the scene. "Wow! That drought really has done a number on this place," he lamented as he closed the screen door to the trailer. "Wonder what that reservoir looks like?" he casually asked as they returned to their preparation for the day. What would they find at the lake?

In a kind of stupor, they completed their breakfast. "Well, why don't we take a walk before we hit the road again?" They had planned to reach the Rockies sometime late that evening. Cleaning up after their meal and securing everything in anticipation of pulling out after their hike, they left the trailer. Walking a short distance, they found an empty campsite (not many had been filled for their night there) and noticed a tall white-barked tree—without leaves. A quick glance and reach for a low limb brought down a very dry twig, for he had extended his arm—a habit of many years—and had grabbed the limb. It snapped between his fingers without any effort—not like green twigs of a healthy tree. "My word! It's dead!"

"This is the exact spot where we stayed that first time," she said. "I know it is because the tree is on the side of the tent site," and pointing to a large boulder adjacent to the tree, "that rock has a very flat top with one smooth side. It had looked to me like somebody had tried to carve something out of it."

"You're right! And look at this tree! It grew, but it's deader than a doornail." Surveying the area, they saw all the other trees in the same state. "The years we have been gone couldn't have killed all these trees. The drought has gotten them." These were city-dwellers who loved the outdoors and were not strangers to nature. Recovering in a way from the realization of what the drought had done, they looked at one another and almost together said, "Let's go see the lake!"

With that, they hit the trail again and took off for the lake. All this occurred in a very short time, but their walk seemed much longer as they realized what the years and lack of rain had done. The trail took them around the spillway—surrounded by dead underbrush and weeds—to the one side of the dam. Anticipating some waves on a small body of water, they could only gasp at the scene before them. He had noticed the smell of dry leaves as they were setting up the trailer the night before, but there was nothing here. No smell of water from a breeze across the surface of a lake. No stink from dead fish! Nothing! All they saw was a dry lake bed—curled dirt where water had once been. No trees! Nothing!

Prominent was a shriveled wooden dock where someone had driven a few beams into what had been water and had bolted some supports to hold up the decking where fishermen had once walked to their boats. Now the decking was shriveled up, the nails were showing, and the whole dock leaned toward the rising hot sun from obvious lack of water that had once swelled the beams and kept the dock square and in shape. The boats were long gone, and nobody was to be seen. The earthen dam had cracks that started at the bottom and grew wider as they wandered in a haphazard way to the top. Several inches wide in some places, the cracks obviously had grown as the sun beat each month upon what now was only a large mound of earth over what had been an arroyo with a stream running through it. No water! No fish! No boats! What a shock to this couple who had come back to remember: the first beauty of a large reservoir with water lapping on the shore; the small dock to which were tied small rowboats and from which fishermen with their rods cast for bass and catfish and perch and. . . .All gone!

"My word! Look at that!" he said, as he took in the deserted scene and began to comprehend what really had happened. "That climate change is real! Just look at what the drought has done! I can't believe it!" But he had to believe it! In the years since they had been here, the conditions had changed. No more could they dream of fishing or launching a small rowboat or enjoying a cool breeze off the water or hearing children cavorting in the shallow water along the shore. All gone!

The couple clasped hands and just stood there for minutes. Flashing through their minds were what they had once seen, what they had dreamed might be, and what now really was. They had seen that beautiful reservoir, had talked of coming back to maybe spend a day by the water. No more! They just shook their heads as they turned away from the scene and slowly took the dry path back to the campground. They passed the weeds, the dry bushes, and the dead trees. Returning to their campsite and the rig, the couple just mechanically took their places in the van.

They said nothing to one another for a long while, as he started the engine, turned onto the road taking them out of the campground. There was no need to drive up past the reservoir. They found the highway and were on their way to the Rockies and hopefully greener campsites. As she looked ahead at the highway and their remaining days of this trip, she ironically remarked, "Well, I'll never have to worry about camping at that site again, *below the dam*."

13

The Vehicle

Setting California. Twenty-First Century

The vehicle—what to do now? That was the question that Jan faced as she contemplated getting rid of her old clunker. The mechanic had said that it needed extensive work—including a new catalytic converter and other engine work that faded into insignificance as he dropped on her the total cost. "Oh my! That can be a significant down payment on a new vehicle," she thought as the details of the repairs were smothered by the shock of the cost.

She had only dropped her car off at the mechanic before work that morning because the thing was making "some funny sounds under the hood." From her mechanically minded mother, she knew what it could be but tried to put that out of her mind. The repair shop had given her a ride to work, and she had gotten a lift from a co-worker back to the shop in the evening. That's when the bomb of cost for repairs was dropped. She thanked the mechanic for the estimate and passively responded, "I'll have to think about that."

"I fully understand, ma'm. This here is a big charge—even I must admit that. Of course, our work is guaranteed for three months."

"And what guarantee is there that there will not be other problems?" she asked with a slightly frustrated tone.

"Well, ma'm, we do stand behind our work, but you know how it is with an older car. They do have their problems." The mechanic projected a slightly macho male air toward this attractive lady, and that only fueled Jan's frustration and put her on guard.

This budget-conscious driver thought for a moment and looked at the list of repairs—labor and parts. She knew what it all meant and wasn't about to give this mechanic the impression that she was just "any female" that said "yes" right away to an "experienced" male. "Well, thank you again for your analysis. I'll be in touch with you."

With that she took the keys, gave the mechanic a smile, and headed for her clunker. It started and ran, but that's about all she could say for it. She had some real decisions to make. "Shall I repair this thing, buy a second-hand car and possibly somebody else's problems, or go for a new one and the immediate loss in value as I drive it out of the dealership's parking lot?" Then there was the thought about car or pickup.

She did love camping and being out in the forest. From her parents, she had gained a real appreciation of the outdoors.

However, Jan had been dubbed a "tree-hugger" by some college classmates who spread their derogatory palaver wherever they could. "Yes, indeed, I do love trees and the environment! Saving this planet for the next generations and doing something about global warming are important," she had told them. She could go on and on until those "jerks shut up," as she had described them to her mom back home on a Christmas break while still in college.

"But back to now," she thought as she forgot those critical skeptics. An environmentalist, yes, but she was facing a decision about how she might get to those trees for camping and, of course, commuting to and from work. She had to be one of those people—a commuter—because public transportation was not convenient or, for that matter, even available where she lived.

"The first thing I'm going to do is some research on the WEB," Jan thought as she managed to get back to her apartment with her clunker. Taking some fresh vegetables from the refrigerator, she put together a salad and found an apple for dessert. "Now's the time for me to make a significant difference in my carbon footprint," she thought as she was still mulling over which way to go. Realizing the chances that she would take in trying to repair her car—plus the repair costs that she could not afford nor finance—she finally was contemplating a good used car or something new. But there was that lingering thought about camping—a small trailer in her future? "No," she thought, "I'll keep using my tent and the flashlight—hiking gear that's worked well."

That camping experience was a treasured way to get away from her work and apartment and out into the open. She loved to visit a nearby National Monument for an overnight and some hiking. Stories about forest campgrounds and no restrictions on guns made her nervous—besides there had been some newspaper stories about a couple of guys who had made trouble for two women up north in a forest preserve. She trusted the National Parks but carried her mace. Pepper spray, also, and an old billy club from her father's police days provided reassurance to this working-out, in-shape young woman. And then there were a couple of private RV parks that she knew about and had security. So, the tent would do and that meant no pickup and a car, instead. Besides, some pickups still got only twelve miles per gallon with a carbon footprint of nine tons—nearly five times as much as a high-efficiency hybrid car. "Buying a polluter could only add to the carbon in the atmosphere and contribute to the earth's rising heat," she thought. "It's a big decision—not just for my pocketbook but for the environment."

Tonight, mulling over all those personal convictions and financial commitments, Jan stopped her meandering thoughts and turned to her computer.

What could she find on the WEB that would help? Categories like sedan, sports car, two-door, color—those and a number of other options were the first decision. Then, there were Blue Book prices and even a consumer-magazine study on efficiency and several other comparative reports on makes and models. American or foreign? Electric, standard combustion, or hybrid? A myriad of possibilities started to surface.

Then there still was that business of new or used? In this latter debate, she remembered the experience of an uncle who had a second-hand car put onto a shop's lift and found that the car had been in an accident—and nobody told him. That left a lasting impression on her, and this time she thought, "Why can't I go new, have the ability to put it on term, and then budget the payments?" That decision was made, but what kind of "car"? Again, she remembered the carbon footprint and checked the WEB. A convenient and quick site came up on the search, and this environmentally convinced consumer decided on a hybrid. "No doubt about it!" It had taken only a moment, and this was the real convincer. "That hybrid will get 50 miles per gallon and makes a carbon footprint that's only about a fifth of a pickup. And, besides, it'll carry all my camping gear." She had been taken on a date with a fellow who owned a hybrid and had lauded all its strengths. "Maybe the electric car will be for me someday, but those limited distances of 80 miles do not begin to reach where I need to go with work and recreation."

All right, the decision was made. Jan had a restful sleep that night, for she even had decided on going with a manufacturer who had, for years, been turning out hybrids. "They had all the 'kinks' knocked out" to quote—in her mind—what her mom had said often. Rising early that Friday morning, she called the mechanic, thanked him for the work, and told him what she had decided. That evening after work, she went to a local dealership, pushed past the first guy who came to her car when she parked in front of the showroom, and found another salesperson—a woman—to help her.

"I want a hybrid," and from there she outlined the specs. "This car will make a minimal carbon footprint. I'm trying to reduce global warming so that we may be able to 'turn down the heat', at least a little bit," she shared with the dealer rep, who positively responded with her own list of reasons to buy "environmentally friendly." She had made her decision. No longer did she have a debate about *the vehicle.*

14

The Big Trees

Setting California: Sequoia National Park, San Joaquin Valley. Twenty-First Century

The big trees—she stood looking up at them—awe-struck by their majesty. The massive trunks climbing to the distant branches high above the ground—all were outlined against the blue sky with cumulus clouds floating by. She felt as though she were moving even as she stood enraptured at the base of one of the biggest. The impression was breathtaking; and as she rubbed her hand over the bark, it felt as though she were rubbing a furry animal. Transfixed for the moment, she momentarily forgot what had brought her here.

Grace had just come from Des Moines on a cross-country trip that she had promised herself a while ago. She had done it alone, at her age, despite warnings from family. "I promise that I'll be in touch every day." Well, she had seen a *tornado* north of I-80 as she came across Nebraska—a big dude. Followed by that was a *downpour* with lightning and *golfball-sized hail.* She had managed to find an overpass under which she scooted just before the large hail started—thereby saving her car. Having crossed a major creek earlier on the Interstate, she saw the stream flooding nearly over the bridge with water spreading into nearby fields. But that was all several days ago. At the moment, she stood at the base of this gigantic Sequoia, while still staring up into the azure sky.

"They are magnificent. I've never seen anything like it," she thought as she promised herself a report back to family in the Mid-West. Realizing a stiff neck, she finally transferred her gaze down the trunk to the tree's immense base. "Right now I'd better get back to the car to find the campground and a suitable place for my tent." She was an outdoors person, through and through, and staying in nature under the protection of the National Park's service was her idea of roughing it. Reluctantly finding her car, she checked the site map provided at the entrance to the Park and found her way to the campground. Traversing the major roads through the campground—as was her habit—Grace found an unclaimed spot by the stream. It was beautiful, and this traveler quickly returned to the store and made her reservation.

Realizing that she still had some sun left, Grace drove a short distance from there and found herself on the Generals Highway. She located the cutoff for which she was looking and quickly took a spot in the parking lot for Moro Rock. With still ample evening sun, she took the quarter-mile trail to Moro Rock—over 350 steps—and climbed out onto the Rock. Looking west through a canyon to the distant valley below, she recalled having found the San Joaquin Valley on a map back home. Now she had just driven through it and up a steep grade into the foothills for a magnificent view of the Park's big trees at the top.

As she now gazed west toward the setting sun, she couldn't forget what she had seen coming through that valley. What had been orchards were now shriveled piles of sticks where fruit trees had once stood. A caterpillar was in the field pushing the trees down and into piles. Workers were destroying them because there was no water—*drought.* At this distance and in the evening sun, all that devastation was obliterated, but the memory was seared in her mind. It was as though a pall had been placed over the distant valley while the brilliant setting sun shown over the cover with hope for a new tomorrow. But how would that day come with the drought so fierce?

Reports had come to the Mid-West about the possibilities of scarcity of fruit and nuts. Certainly the rising prices in her most recent visits to the grocery store attested to shortages. Now she had seen the source of the problem—no water to irrigate the trees. Years of tender planting and pruning—down the tube. Days of harvesting—no longer occurring. She could only imagine the financial crises created for the owners, pickers, marketers, and the host of other processors involved in providing the fruit and nuts to the nation. That was a bad nightmare; and waking to the moment, she cast her eyes to the sun's final brush strokes of pink and faint yellow on the distant clouds. "This day is done, and I have taken every light hour out of it. I must get back to the campground."

Gingerly walking on the trail back to the parking lot, she found her car and with a shiver hopped in and locked the doors—as though a specter from the night would grab her. Certainly she had been enthralled by the beauty of this place, but she noticed a slight yellow glow to the north—certainly away from the sunset that she had just experienced. On her coming into the Park, she had seen some smoke in the distance, but thought nothing of it. For a moment she had put the drought and its effect on the valley behind her, but now suddenly she was confronted once more with its effect. "Is that a fire? Goodness, I hope not. That would be the final blow to what I had dreamed would be a wonderful trip."

Thinking it important to know what was occurring, she made her way back to the campground and found some coldcuts that she had iced the day before. Tonight on some freshly cut bread, the meat tasted so very good, and the meal satisfied what had been growing hunger pangs. She took a couple of swallows from an opened wine bottle and finished off with a crisp apple. With a battery-powered lamp lighting the nearby picnic table, she finished putting up her tent and throwing her sleeping bag and pillow into her tent's inviting opening. But then she remembered the ranger's note regarding a talk, checked her watch, and decided she had just enough time to make the 8:30 presentation. She had always enjoyed ranger talks, and tonight she could learn something about that slight yellow glow to the north.

Walking with flashlight in hand to the ranger talk, Grace spotted the lighted marker beside the amphitheater where the presentation was to be. She found other campers buzzing about the dim light to the north. The ranger had a small crowd of campers gathered around him. Grace quietly approached the group and heard the ranger talking about a fire that had started the previous day. "The prevailing winds are carrying the fire north, away from us, for the moment. This is a real reminder to all of us to avoid any camp fires and to put out any smoking materials. The drought is severe," the ranger reminded already very conscientious campers.

"Once more a reminder, even here, of the seriousness of the drought," Grace thought as she listened attentively. Notices, of course, were everywhere in the Park about the seriousness of the drought and the danger of fire. Stories were around about how fire could be good for Sequoias, but those same people had not been around in tinder-dry conditions that could allow fires to roar through the forest and to destroy these precious trees. The ranger wrapped up her brief conversation with the smaller group and gave direction for the entire group to be seated. Coincidentally the lecture and slides for the evening were on the drought and what effect it was having upon the vegetation and wildlife. "How often have you asked yourself what happens to the animals of the forest when a fierce fire marches through the trees? They can be the forgotten creatures that suffer death and burns as a result of the fierce heat," the ranger explained to an empathetic audience. Pictures showed the shock to the smaller plants, the underbrush. Concluding her remarks, the ranger assured the group, "Someone will come to your campsite if the wind changes and there is a real threat. We will have someone guide you out a safe way." As the ranger spoke, Grace remembered the story of a friend who had been on the North Rim of the Grand Canyon when some foresters had set an underbrush fire which had gotten out of control—her friend and all the visitors followed a ranger's vehicle north as they were ushered away from the rim and the severe fire that had resulted from the out-of-control "set fire." Returning to the ranger's assurances of warning, Grace said goodnight to the campers around her and took the path back to her tent. The smell of the trees, the moan of a breeze wafting through tops of the tall oldsters standing around her—this brought her back to why she was here—an appreciation, once more, of nature and its beauty. Tonight she would sleep restfully—hopefully—beside her flowing stream—the spot that, by some chance, had remained empty and she was able to claim.

Awaking the next morning, Grace rolled over onto her back and, staring at the wall of her tent while the early morning sunlight highlighted the top, seriously asked herself, "Is this a trip that a retired professor should be on alone?" No one had come to her campsite in the night; so presumably things were all right—for the moment! To this retired university geology professor, the threat of the distant forest fire brought to mind, once again, all her memories about global warming—excessive fires, drought, pouring rain with flooding, stronger hurricanes, fiercer tornadoes, large hail—some of which she had experienced coming across the country on this trip. Now in the Southwest, she saw the threats to the beauty that momentarily is and what had been. But, she was in the forest of Sequoias at the moment, and she overheard nearby campers talking about a ranger who had indicated that the fire was still moving north—"away from our trees." With that assurance, Grace quickly dressed for the day, had a bite of breakfast—ice-packed orange juice and cereal with cool milk—and headed to the Visitors' Center.

There she found a typed report on a bulletin board—no wi-fi or computer or anything else for her at this moment—she stood with a couple as they read the news. Just like the rumor that she had heard back in the campground, the fire was moving northward away from the Sequoias. Having read the notice, she continued to the Visitors' Center. Entering the gift shop, she found a book on the trees and purchased it. Seeing that they had a presentation in the theatre on the Park's history, she attended the program and then returned to her campsite. This day would have time for another more extended hike through the big trees—including around the General Sherman, the largest living tree in the world.

Her three nights of camping passed quickly, but she was determined to return down through the San Joaquin Valley to learn more of the plight of the growers. She would, indeed, go to the coast and spend at least a week at the beach and drive along the coast—after all, the West Coast has some of the most beautiful scenery in the nation, according to many accounts. But now packing up her camping gear, she reluctantly entered her car to bid farewell to these magnificent sentinels from the past. She would not live to know how they survived this drought, but she did know that they had survived many before this.

Leaving the Park and continuing her journey through the Valley, she stopped for gas, and at the station struck up a conversation with one of the locals who also had come for gas and a few groceries. The tragedy was too apparent as the grower responded to her questions about the effect of the drought. After the brief encounter, she went and found more details in a newspaper carrying a banner "Drought Takes Its Toll." Making the purchase, she tucked the paper under her arm and left the store.

Once in her car, she just could not rest until she had at least perused the front-page story. She left the pump and parked her car on a corner of the station's lot. There she returned to the paper. The article noted that three decades ago a well could be less than three-hundred feet deep and irrigate citrus, pistachio, and almond trees. Now there was no water for irrigation of the San Joaquin Valley. A local well-driller was booking six months out to drill wells deeper with the hope that water could be found. In the meantime, the article noted, many growers were cutting down their dead trees and hanging onto the land with the hope that conditions would change. With less snow in the Sierras and growing demands for water from urban areas, the orchards were dying.

Having completed skimming the article, Grace stopped for a moment, reached for her phone and placed a call—something she had not done in a couple of days. The voice at the other end seemed almost in another country. She reassured her sister back in Des Moines that she was fine and so appreciating California's natural beauty. "I will have so much to tell you about the drought when I get home, but now I am leaving the Sequoias for the coast. More later!"

With that completed, Grace turned over the engine once more and headed west. She was going to the coast—Los Angeles—north and south. Her experiences would be new there, but she would never forget the beauty and the potential plight of *the big trees.*

15

Ice Age

Setting Southeastern Arizona: Interstate 10. June, 2014

"*Ice age*—that's what it looks like—the results of a lot of grinding and moving," Jeremy thought, as he looked with camera in hand at the pile of giant granite rocks. Yet, the literature from the last stop said differently. He had just stopped in the last town for gasoline, and the attendant said that it was record heat for this day—114 degrees in the shade. "And here I am, out in this sun, taking a picture of that pile of rock. I must be nuts." Now realizing the extent of the heat, he wondered if his seven-year-old SUV really could make the trip—even with all the maintenance before he left.

Standing here in southeastern Arizona at the base of that mountain of boulders and seeing the desert around, he wondered, "How could one not imagine that this all was the work of an ice age with giant glaciers grinding these boulders and bringing them this far south?" But then he remembered from his college geology class that the ice sheets had not come this far in the last Ice Age, only across the continent as far south as Wisconsin. "But, wow, these certainly look like what has been left at the bottom of a glacier," he countered in his mind.

Studying that literature from a visitors' center, he had learned that the rocks were the result of pressure and broiling up from the interior of the earth. Over millions of years, erosion had exposed this granite to the atmosphere that had started to shape the rock. With exposure to water and air together with cracking from cooling, they had taken these rounded shapes.

"Well, that's all very interesting, but here I am standing in this furnace," Jeremy thought as he felt the heat through his shoe soles. He was taking a tour of the Southwest—at the beginning of the summer—but it was the only time he had, since he was now teaching high school English in a small Wisconsin town and had only summer breaks. "This is the summer that I am going to see the Southwest," he had mused. This teacher had decided that San Diego would be his destination, with stops in Denver and Albuquerque, from where he eventually would head through some mountains in southwestern New Mexico to pick up I-10. And those plans were what had brought him to these boulders in what is called Texas Canyon in Arizona.

Now he had yet to travel across the width of the states of Arizona and California. Originally he hadn't really given the heat a lot of thought, since he would wake up each day to cool mornings back home. With friends, he had talked of climate change and earth warming in making these plans, but that was more academic than really actual—so he had thought. Now he was in the middle of it all—nothing but scrub vegetation, hot desert and barren mountains all around. "Why did I do this? Then there's the old SUV, to boot."

"Oh, well, I've made a lot of progress and am really on the way now. In another day or two, I will be enjoying the *cool* beach on Coronado Island in San Diego." He quickly finished his photographing and headed for the SUV. Relieved that it started, Jeremy moved from the rest area and merged onto the Interstate. Invigorated from his prior night's rest, he decided to push through. Entering Tucson, he stopped for gas, noted the heat, and again hit the road. Exiting to Interstate 8, he felt relieved that he was missing the congestion of another metro area by taking this route south of Phoenix. Still concerned about his vehicle's stamina, he was seeing desert all around him and wondered, only for a moment, what he would do if his SUV would give him serious trouble. "But I did have it checked and serviced before I left home," he reassured himself. By the map, he would have four-thousand-foot mountains to traverse before he arrived at San Diego. Back in the present, he observed the desert's beauty to the south with volcanic peaks reflecting the late afternoon sun—but it still had a foreboding air, if there were trouble.

One more gas stop in a desert town that was just at sea level before he hit those mountains—what a relief to be this close. Twilight provided him an outline of the distant rise from the desert floor. As he reached them and started gaining altitude, he noted once more, "Doggone it, there are those boulders again—this time whole mountains of them!" Several times, the highway descended a couple thousand feet and then rose to another pass. The highway had little traffic, and with the sun completely set, his headlights were the only guide for him as the roadway wound up and down through the mountains. Then, he saw the aura of San Diego and had a sense of relief.

Gone now were all those academic thoughts about an ice age, although that desert had made him wish more than once he would have had some ice with which to cool off. The SUV's air-conditioner was working, but he still had felt the heat through its, at times, feeble attempts to cool. Oh, well, he was in San Diego now and found his hotel right off the highway—in the complex of motels and hotels just before coming to Interstate 5.

This relieved traveler parked, grabbed his bag, and headed for the check-in desk. A quick confirmation of his reservation, and he was signed in and headed to a welcome shower and bed. Sleeping late the next morning, he rose to morning fog, for he was just at the end of what the locals called "Gloomy June." Having a leisure breakfast, he stuck his nose outside and found—to his great relief—respite from that record heat of the desert. Returning to his room, he donned a swimsuit and headed for his SUV and then Coronado Island.

The desk attendant had provided some pointers on how to find the beach and where to park. Negotiating the Coronado Bridge, high over the harbor, he came into a rather busy urban collection of homes, shops, businesses, and restaurants. Finding his way to the vicinity of the island's major hotel and then the beach, Jeremy parked and was in seventh heaven for having found the cool Pacific and "maybe even," he thought, "an attractive, intelligent woman with whom I can become acquainted." The journey through the desert was history, and this traveler left thoughts, for another day, about the return trip and the mistaken assumptions of the *ice age*.

16

Highway 66

Setting Southwest, Highway 66. Twenty-First Century

Highway 66—that's what the couple had planned on tracing in travels from their Missouri home to Los Angeles. They had taken Route 66 when it was only a four-lane highway back in the '50's in Illinois. Right after WWII that Illinois stretch had gone to four lanes from Chicago to the Mississippi River. But then the Interstates took over—including Interstate 55 between Chicago and St. Louis. Nostalgia, however, preserved sections of the old highway, and they planned to reminisce by taking a sentimental trip—this time from their home near Springfield all the way to the West Coast. At the moment they were at what had been, and yet remained, the highest point in the journey—Flagstaff, Arizona.

Maggie and Matt were reliving their youth and former travels. The couple had lived in Chicago all their working years and had retired to the Ozark Mountains soon after "laying it up," as Maggie frequently said. They were enjoying living near Branson—just down the highway from Springfield—Missouri, that is, for Springfield, Illinois, was also on the old 66 route. There were many jokes about mix-ups from travelers who headed to the wrong one. Picking up their nostalgic trail in Springfield, they had ventured out in their dated, but repaired, sedan with some trepidation about how reliable the car and their own bodies, for that matter, were for a trip of over a thousand miles to the Coast.

Great and exciting plans had been made just before they had retired in Chicago. Travels with Missouri relatives to the Lake of the Ozarks and southward had convinced them that this would be the real place for their retirement home. They had found a plot of land on which a modest home was built near Springfield. "We're still going to have four seasons, but winters won't be as harsh as northern Illinois," Matt had rationalized. After the move, they settled in and really felt "at home with the local folks," they wrote back to Chicago. Restaurants and some of the theaters in Branson were an attraction. "We get to see Andy Williams and Dolly Parton and many other entertainers whom we have known in our lives," they noted also in correspondence.

It was a "dream come true," to quote some of their contemporaries, until . . . . "Yeah, until the Great Recession hit and then it was 'hail, Columbia',"

Maggie lamented. Part of their retirement investments had been in the stock market, especially, contractors and building products. Home prices plummeted, as did their savings. So, it was not so bright, and "things had to last"—including their car that had been new when they moved from Chicago. But after the crash—and they thought it wasn't too different from the '30's Great Depression with what they had suffered—things started to come back, *slowly*.

They started to recover; and after scrimping and saving, one day Maggie said, "We need a break—a change! How about a trip to the Coast?" They had managed to visit St. Louis, but "that was just up the road," and this wife of many years wanted more. So, the couple planned and checked and used their AAA membership to set up the trip—Springfield to L.A.

The two were going to take a nostalgic trip and retrace some of the routes that they had taken as a young couple. In planning, they remembered how Maggie's uncle and aunt had left Oklahoma in '33, packed up the few sticks of furniture that they could load onto a Model A truck and had made the arduous trek to California. Before Highway 66, there had not been a route all the way from the Mid-West to the West Coast. Both of them knew that trek between Chicago and St. Louis like the backs of their hands. After all, they had worked in the Windy City all those years and had traveled many times to see their son who was working in suburban St. Louis. The start of the trip was routine—having the auto checked thoroughly, new tires, packing—all the things that this couple had known over their over forty-year marriage. They had "launched" from Springfield and had spent some time in Oklahoma—exploring some of the shops along the way as well as visiting with locals who shared all kinds of folklore with them.

Now they were in Flagstaff. Reservations had been made in advance to stay in a motel along the old route through "Flag." Taking the first available exit, they wound their way through the east part of the city and found the AAA-recommended motel that obviously had been around for a number of years. Checking in and settling into their room, Matt was the first to suggest it—food! "I'm famished. Let's find a restaurant!"

"That's fine with me! The desk clerk had mentioned that locally popular one just down the street."

They returned to their now road-dirty, but still perking, car and headed for the restaurant. Its blinking neon sign—for by now it was getting dark—reminded them of years ago, when without kids, they had found their way to a similar, family-owned café with home-cooked, delicious food. Even in that short distance, Maggie was musing over their leaving the Interstate and reminded of their one exit back in Oklahoma.

They had taken one of the exits off the Interstate and had found an old stretch of the highway. "Let's stop for a moment," Maggie remembered requesting as she had scanned the horizon. Matt had pulled off to the edge of the old highway. Both travelers had gotten out, and just drank in the western horizon—the farmland and fields of sunflowers. Down the road, they had found some small-town shops and had enjoyed chatting with the local folk and hearing their stories.

She was awakened from her musing as they pulled into the cafe's parking lot. "Matt, do you remember our stop in Oklahoma and my saying, 'I can almost hear those old flivvers and jalopies rattling across that highway on their way to California'?" No answer, but that didn't stop thoughts about conversations and reading racing through her mind. Many improvements and changes had been made to the route in Oklahoma—including the conversion to turnpikes—the first one from Oklahoma City to Tulsa in '53. Maggie had heard stories about people staying off the turnpike to avoid paying the tolls and sticking with "old 66." She had read about how later a stretch of the super highway was added from Tulsa to the Missouri Border, west of Joplin, and was then picked up as Interstate 44. But she also read in the article that, over the years, old highway 66 was preserved in places and still paralleled parts of the turnpikes.

Their car was parked, Matt was getting hungrier by the moment, but Maggie couldn't stop thinking and talking about that stretch of highway. "Remember how Will Rogers' name was tied to the turnpike? And I just can't stop thinking about those Oakies trying to escape the Dust Bowl and Depression by going to those California farms for vegetable and fruit picking. Just think of it—a drought, that Dust Bowl, and the Depression, to boot—they really suffered." Still recalling, she added, "What hope did they have, except that they might find something better in California? And we're headed that way—gosh, how some things change and others don't."

"Yeah, we're headed that way all right," Matt added, "but we could have some of the same troubles those people had if this car doesn't make it—including trying to get across that hot, dry desert!"

"Oh, don't be so pessimistic. We're going to make it! And if this darn car has trouble, we'll just have it fixed." Maggie fussed at Matt with a smile about his always finding the dark side of a cloud. "You know how our families in North Texas suffered from all that—including grasshoppers eating everything in sight."

By now, Matt—having frustratingly watched the crowd going into the restaurant—got out of the car. "Let's get some supper," and with that Matt put an end to the conversation—for the moment. They entered the friendly restaurant with frilly curtains and a country atmosphere. Seating themselves, they found a table by a window, a smiling server, and a tempting menu. After ordering and having their meal served, Matt couldn't wait and just devoured his dinner.

Finally slowing down with a dessert (a shared piece of apple pie a-la-mode) and a cup of coffee, he noticed a friendly couple across the aisle. He struck up a conversation and, in the usual tourist fashion, explained, "We're here to see the sights and reminisce—in part to repeat a trip of many years ago. We're very interested in old Route 66."

"Well, welcome!" the stranger responded. "Have you seen the shop just down the street? It has a lot of history and souvenirs about the Highway. We live here and know a number of people who have found any number of 66 souvenirs there."

"We'll have to check that out tomorrow." Maggie joined in the conversation. Exchanging a few more convivialities, the couple excused themselves. "Enjoy your trip!"

"Thanks. We appreciate the info," Maggie added. She and Matt finished off their meal with a second cup of coffee and paid their check. Back at the motel, two weary travelers showered and went right to bed. The next morning, they got up early, had a continental breakfast at their motel, and decided to visit the shop, which also turned out to be a visitors' center. The couple figured that walking would be better for them than driving. "We'd been on the road too long, and needed the walk!" they later wrote to friends back home.

In the store they did find a couple of brochures and a Route 66 sign that they couldn't resist buying. "We have a lot of stuff at home already, but we just have to get this," Maggie suggested to Matt. Making their purchase, they found a coffee shop next door and sat down to a hot cup of coffee and a couple of donuts. "After all, we're on vacation," Matt rationalized, even though he knew the walk had been designed to work off calories that he should be counting—especially on travel.

The couple began to peruse the materials that they had gathered from the center. "Oh, look at this. This brochure notes how President Eisenhower in '56 signed the Interstate Highway Act, which began the slow-but-sure reduction in use of Route 66. We saw that in some of our literature at home. It says here that the Route finally was decommissioned in '85. And look at how the Interstate took over the old highway," Maggie noted as she pointed to a chart.

Substitute Routes under Interstate System

Interstate 55	Chicago to St. Louis
Interstate 44	St. Louis to Oklahoma City
Interstate 40	Oklahoma City to Barstow
Interstate 15	Barstow to San Bernadino
Interstate 10	Through Los Angeles to Santa Monica

Maggie read that the highway had been in existence over fifty years, and that the Pasadena Freeway that they would take in Los Angeles was part of the old 66. The brochure also noted how the highway had originated in '26 and was completely paved by '38. "Chicago to Los Angeles—that's 2400 miles," Maggie read from the brochure.

"Remember how Eisenhower traveled Europe during WWII and finally saw the German Autobahn? That thing had been built for fast travel to transport military equipment across Germany. The President even wrote about how trying his more-than-twenty-day trip across the U. S. had been as an army officer before WWII! My word, it was no wonder that things had to change!" Matt blurted out, with his not-too-unusual way of getting launched on something.

"Yeah, and here we are enjoying the results of his push to upgrade our nation's highways," Maggie added. They continued to look through the brochures and nostalgically remember the history that had been part of their lives. They learned how Flagstaff had designated one street as Highway 66. Having traveled the whole route at various points in their lives, the material brought to mind their plans to make this trip. They knew the whole route—from Chicago through Illinois, Missouri, Oklahoma, the Texas Panhandle, New Mexico, Arizona, and Southern California to L.A. Now they were on the last stretch. Having spent nearly an hour perusing the materials and with lunch approaching, they each ordered a small sandwich. After a second cup of coffee, they gathered up their materials and headed back to the motel.

"I think that I had better have the car checked at that fast-lube place across the street from our motel. You can enjoy the pool while I go over there and have them change the oil, check tire pressure, and so on." Matt just didn't quite trust their old car—even though it had done beautifully—*so far*! There was some of that pessimism coming out again. So, back at the motel, they went their separate ways—Maggie to the pool and Matt to the fast-lube.

An hour later, Matt returned to find Maggie at the pool. "The car was fine, and we're ready to go on the last leg of the trip." Matt went back to the room, changed into his swimsuit, and joined his wife again at the pool. As they lolled in the afternoon sun, Matt remembered the news the prior evening. "The TV report last night from Phoenix was not good. Just too hot!" they both agreed, and rehearsing their next move, they agreed to stay with I-40 instead of heading to Phoenix.

With record temperatures in Phoenix, for this trip they would avoid the Valley of the Sun altogether. Besides, there was the car and how it would take the heat. They both couldn't forget how dry the Mid-West had been over the past several years, and hearing about the desert conditions in the Southwest just made them want to "enjoy the Pacific breezes." Now they were getting very close. "I heard from one of the women here at the pool that Williams has a lot of nostalgia from Highway 66 days. How about stopping there tomorrow?" Maggie suggested. "I hear they have at least one museum and, being on the old route, have worked to preserve the history."

"Sure. Let's go there! We're all set, and this break in Flag has been a real respite from the trip." They both were feeling their years and welcomed some time off the road. So, the next morning—with the car's engine running smoothly—they left Flagstaff and went the few miles down the Interstate to Williams and its memories of the old highway. Impressed with the "66" lore around the town, they found a motel and stayed overnight.

The next day, the traveling couple hit I-40 once more and headed west through the desert to Needles, California. As they came off the high country into the desert, they crossed the Colorado River. Studying the river's size, Maggie mused, "Wow, this isn't the Mississippi. That bridge over the Mississippi was really something in comparison to this one. What an improvement that thing must have been when it opened and rerouted traffic north of St. Louis."

"Yeah," Matt chimed in, "that Chain of Rocks Bridge over the Mississippi was an engineering marvel in its day. I remember how we could see the expanse of the river carrying tug boats with barges and even a side—wheeler at times. That's a lot of water, but what we are seeing here is a mighty river in its own right. This river flows for hundreds of miles from

Colorado's mountains through canyons and desert to get to this point."

Having crossed the Colorado River and endured 115-degree temperatures across the Mojave Desert after Needles, they came to Barstow, where their highway dead-ended into I-15. They turned south to Los Angeles, and as they were leaving I-40, Matt recalled, "What a difference this highway is from what people took through here forty years ago! Certainly all those re-routings of old 66 did improve the highway, but look at what we have now!" Wider lanes, improved pavement, double lanes to four lanes—all had enhanced the route; but now the Interstate System had taken over; and this couple appreciated the change!

With all of those highway realignments, the couple had agreed (before they even started the trip) they would avoid most of the now renumbered routes that 66 had taken over the years—including the stretch through Oatman, which was now virtually a ghost town and had been on the old western Arizona route to L.A. until '53. Hairpin curves had made some people afraid of that part of 66. As they had discussed this stretch when they were still back home, Maggie had noted, "Stories were that even some at times asked for a driver to take them through Oatman." But all that was in the past, as they now approached the coast.

In these last couple of hours on the road, the two passed some of the time recalling their new experiences. They hadn't been around for the highway's official start in '26, but had heard of its decommissioning in '85. They once more recalled what they read about the old highway's story over its nearly 60 years of serving the western United States. To enhance their nostalgia, they played some of the old songs that brought back a lot of memories—now nearing the coast, they popped into their disk player, "Do you know the way to San Jose?"

Their memories had taken the couple through the happy experiences in their travels along the route. As they entered L.A., they rolled down the car windows and felt the breeze off the coast. "Goodness, isn't that wonderful?" Maggie exclaimed. The hot desert was a long way off at the moment! Maggie and Matt were at the end of their trip, and looked forward to a hotel and lots of time on the beach. Their car had "made it." Past, for the moment, was their reminiscing about *Highway 66.*

17

Canyon Country

Setting Colorado, Middle Park, and River Country. Twenty-First Century

Canyon Country—that's what they were recalling as they sat before a roaring fire high in Middle Park, southwest of Rocky Mountain National Park. Surrounded by the snowy Rockies in this broad valley, they had visited Grand Lake on the Colorado River in the afternoon. Coming here for another wedding anniversary, they often shared memories of their many years of travel and reminisced about all their experiences. This evening they were remembering the Colorado River and all the beauty that the mighty stream and its canyons had brought them. As they sat on overstuffed chairs in this spacious lobby, they gazed through the broad plate-glass window at the silhouetted mountains with the setting sun beyond them. "Up there are the headwaters," she whispered as she pointed to the northeastern peaks. They had chosen this hotel for its proximity to the river and the breathtaking views of the Rockies.

"What was your most memorable experience in your encounter with the river?" someone blurted, interrupting their contemplation and total absorption in the scene. Breaking the couple's trance, a soiree of strangers had just raucously entered from the bar and was watching the couple.

Aroused from his revery and thinking for only a moment, Rob retorted to the stranger's question: "Nothing really! It was all too beautiful to say there was one thing." With Rob's short response, the partying questioner abandoned the couple for his boisterous friends.

"Nothing really!" Rob considered his answer to the stranger. "Yes, that is true!" After all, they had motored along the Colorado's banks for miles; taken a train with tracks along a stretch of the river; boated lakes now covering several canyons; floated through a canyon with tracks on one side and a highway on the other, past a major tributary; rubber-rafted below a major dam just north of the Grand Canyon; and jet-boated a part with high sedimentary yellow cliffs on both sides. All that travel had names and places, but for this moment Rob's mental flashback took him from the river's Rocky Mountain source to the Gulf of California and back to the lobby.

This moment was theirs, quiet and focused. They had been together many years, and this time was treasured, "for we never know if we will be back," they had frequently said to themselves in the last few years. Bonnie reached over to squeeze his hand, as Rob remained comfortably reclined even as the other guests had moved through and invaded the adjoining dining room for their evening ritual of food and more drinks. The two refocused in their reverie, for a quiet had returned to the room and the sunset's remnants remained only in swatches of pink and yellow on clouds in the distance.

Another day was drawing to a close with just a few left on this visit—they had come from their home in Denver—just across the mountains. The couple had no doubt as to what their activities as hikers and environmentalists would be in the few remaining days here in this beautiful setting. They were very conscious of the reduced snow in the mountains and the drought that had faced the Southwest for more than a decade. The river's headwaters didn't have the deep supply of snow that always had been there, and the high desert through which the river flowed was suffering. The seven-state area served by the great Colorado was facing some monumental challenges. But that was another problem—something that Bonnie had faced frequently in her "working years" as a hydrologist employed by the Bureau of Land Management. But that was then—at this moment Bonnie wanted to focus on their time together. They were "staying a few days," for their income now was "fixed" and limited their travel.

The next morning—after a restful night—at breakfast on the outdoor patio, they met one of their fellow residents from the night before. With some hesitancy the lady approached their table. "I was in that group last night that obviously disturbed the tranquility of the lobby. You were transfixed by that beautiful sunset, and I truly am sorry that we disturbed that moment!"

Realizing the kindness of this apologetic woman, Bonnie with a smile responded. "Oh, thank you very much. No apology necessary! That beautiful lobby is for all residents, and we were just lost in our own thought—appreciating that gorgeous setting." Noticing that she was alone, Bonnie invited the stranger, "If you have no one else, won't you join us here at our table? We always love to visit with fellow vacationers and travelers."

"Well, truth be told, I am traveling alone and would be happy to join you. Thank you!" Obviously trusting the couple with the proffered explanation, the lady added, "By the way, I am Sally."

"Wonderful! We are Bonnie and Rob," Bonnie responded. In gentlemanly fashion, Rob arose and offered their visitor a chair—assisting her all the while. At that moment the attendant came and provided them with menus. "Thank you very much," Rob convivially responded as the diners accepted the embossed leather menus.

They each perused the—what seemed—booklet of options and, after a seemingly thorough study, made their choices. After ordering breakfast, their conversation turned to an exchange about their backgrounds and reasons for being in this high mountain valley.

Politely noting their obvious years, the younger Sally inquisitively observed, "You must have memories galore?" And with that the stories began, for at their age the couple welcomed an opportunity to share their many happy experiences along the river.

"Yes, we do," Bonnie responded with a thoughtful glance across the valley. "We have been to Rocky Mountain National Park often—that's where this great river starts—and we do so treasure it and our years of being on and near it. The source is just the usual trickle as it comes from those majestic mountains."

"Oh, yes, and isn't its flow into Grand Lake just beautiful?" Sally recalled. "My kids and I went camping there and skipped rocks over that placid surface while we watched canoers glide between those tree-covered islands."

"Yes, and Lake Granby is equally beautiful," Bonnie added. About that time, the server came with their food. The aroma in the crisp morning air whet their appetites even more, as they saw the steaming food being served.

"My, isn't that beautiful?" Bonnie commented as she was served. "Thank you so very much!"

The attendant looked with a certain shock—hearing a note of gratitude from what frequently were diners oblivious of everyone but themselves and their own needs. "Why, thank you," came the response from the attendant.

"We were just talking about this beautiful setting and Grand Lake and Lake Granby. What has been your best moment on the Colorado?" Bonnie queried as she looked with grateful eyes to the server.

"Oh, I love it all—that's why I work here seasonally! But, I guess, if I really think about it, the ride on the river out of Kremmling was most exciting. The river moves rather placidly through the mountain park and drops into one of its many canyons. The trip through Gore Canyon on the river is considered one of the most advanced in the U.S. I have been told that some kayakers' number for that stretch is possibly a 7 out of 10—what a ride!" The attendant had momentarily been absorbed in these appreciative visitors and suddenly recalled, "Oh, I had better return to the other tables. Thanks for the opportunity to recall. Is there anything more that I can get for you?" he asked with obvious attentiveness.

"Nothing for the moment. Thank you again!" As they turned to their food, Bonnie noted, "I love to engage these servers in conversation. One can learn so much about them and the area!" The three diners, with obvious anticipation, turned to their food and momentarily and quietly enjoyed the meal. The attendant, returning, again inquired of their needs, if any.

"We're just fine! We will take a refill on coffee and then the check," Rob responded. Sally made an effort to pay; but Rob, with an "I've-got-it-covered" gesture, took charge of the check. With that paid and pointing to a protected area beside a fire pit, Rob suggested, "How about taking our coffee over there to the corner?" The morning temperature had a briskness, and a slight breeze had begun to blow across the patio as the sun's rays created shadows from the trees adjoining the dining area. The cool air, wafting over the patio, brought the aroma of mountain evergreen to the summer diners. They settled into comfortable chairs near the fire—still able to see the surrounding trees and mountains.

"That air reminds me so much of our experience in Glenwood Canyon," Bonnie noted. "Our server's account of the canyon out of Kremling brought back many memories of our travels along the river from Denver. We often have traveled through Glenwood Canyon on the highway, but one time in Denver we boarded a train that followed a stretch of the river a few miles before it enters Glenwood Canyon. Rafters were doing their usual thing as the train passed, but we were more intent on the white water, that canyon, and the trees climbing up the slope. That trip took us to Glenwood Springs, where we spent the night before our return to Denver the next day. But our favorite is to drive through Glenwood Canyon and stop along the way to watch the water and appreciate the aroma of the trees wafting across the rest area.

These evergreens and their 'piney' smell just transpose me to that spot in the Canyon."

"Oh, yes, and have you been in the Glenwood Springs park along the Roaring Fork?" Sally asked as she picked up with her own memories. "I was there several years ago, and the meeting of that snow-fed river with the Colorado is quite something. In fact, I remember my being with a friend, and we were exploring that river valley up beyond Aspen and went up to Marble, where some of the quarrying for the Lincoln Memorial in D.C. was done. White marble lay all over the ground under the trees just below that quarry. On the way back to Glenwood Springs, we ran into a thunderstorm that had dumped several inches of rain on the mountainside. All that came roaring down in a flood with debris and blocked the road. Of course, we had to wait for the road to be cleared."

"Wow, what an adventure that was!" Rob responded. "The River settles down west of Glenwood and meanders through beautiful valleys. Interstate 70 crosses it in several places. I remember when that highway was single-lane, and it took seven hours to go from Denver to Grand Junction—obviously the confluence of the Gunnison and Colorado Rivers. We have made a fall trip a number of times to get fruit from the Grand Valley. When the river leaves that broad, lush valley with its orchards, truck farms, and vineyards, it almost falls into the Utah canyon country. Have you been to Moab, Sally?"

"Oh, yes, and that whole area presents more beautiful vistas and experiences with the river. I've taken a jet-boat into Canyonlands National Park—what an experience that was!"

"So have we!" Bonnie jumped in. "We also, as I'm sure you did, saw it from the overlook at Dead Horse Point. That's all in the vicinity of Arches National Park. Then, south of Moab is another overlook, Island in the Sky, where the river makes a number of sharp turns and eventually meets the Green River coming down from Wyoming. Erosion has certainly done a number on that country! And from the studies of the geologic and climatic changes, we all know how lush the country once was—with forests, lakes, and probably even some swamps. That 'ain't' the way it is now! The current lack of rain and heavy snow in the mountains is taking a heavy toll even on the junipers that have thrived in some of the higher altitudes. Oh, my, now I'm launched, but I can get so excited about this beautiful country."

"Yes, me, too!" Sally agreed. "And undoubtedly you have been to Lake Powell?" With a big nod of heads from both Rob and Bonnie, Sally was launched herself. "I took a boat ride from Bullfrog Marina and spent a whole day on the water with a captain and another couple. He took us into the canyons partially filled by the lake and especially down to Rainbow Bridge."

"We've been to the same place," Rob interrupted. "In fact, we also took a tour boat from just above the dam and went to the same place. Rainbow Bridge is magnificent. With the drought the water level has been falling, and we had to walk a good mile to get to the arch from the dock. But then, we went to Page after our trip on the lake. Have you been there?"

"Oh, of course," Sally responded. "You probably took the float trip below the Glen Canyon Dam? And saw the petroglyphs, too?" Again with a nod of agreement from the couple, Sally related how she had made reservations in Page and was able to go to the same spot where the Grand Canyon trips are launched. All three had made that trip at some point. "And then there's the best of all—the Grand Canyon." Rob and Bonnie had driven Highway 89 south from Page, over some rather barren country to the eastern entrance of the Canyon via Cameron.

"Oh, yes, and we've made the trip to the Canyon several times—as I'm sure you have," Rob interrupted. With a nod from Sally, he asked, "Have you been on a raft trip down the River, Sally?" By now all three of these travelers were mentally transferred to the Grand Canyon.

"No, and it has always been something that I wanted to do," Sally added as she picked up the thought. "I particularly would have enjoyed seeing the confluence of the Little Colorado and the Colorado Rivers on the east side of the National Park. I understand a spring on the Little Colorado turns the water blue before it gets to the muddy Colorado, and I would love to be there. However, the overlook at

Desert View on the South Rim provides a spectacular view of the wanderings that the canyon of the Little Colorado makes as it joins the Big River. Oh, well, we may not be able to be down at the confluence, but that overlook at Desert View has satisfied me! One thing I did do—I hiked a part of Bright Angel Trail from the South Rim. It was the middle of the summer, and I never got down to Phantom Ranch, where it was 106 degrees."

"Wow!" Bonnie responded. "We didn't do that, but we did have lunch at a restaurant overlooking the head of the trail on the South Rim. That reminds me—have you been on the North Rim? There's a restaurant with a huge plate-glass window overlooking the Canyon. We had an evening meal there just as the sun was setting—what a breath-taking view!"

"No, I haven't done that," Sally offered. "It's something I've put on my bucket list! Another part of that list for me will be a visit to the Havasupi village in the Canyon. Pictures of the Havasu Falls and pools there are just magnificent!" Their conversation continued with a kind of youthful exuberance—they enjoyed one another as fellow travelers who had been along the Colorado River and loved it! They talked of its place as one of the seven natural wonders of the world and the debates about its origin. The three had lost track of time, but they continued their reverie. "Then there's the whole route of the river beyond the Grand Canyon," Bonnie added.

"I'm sure we all have been to Vegas and Lake Mead?" Rob asked rhetorically. "Boulder Dam backs up all that water on the Colorado for generating hydroelectric power and supplying Vegas with water. But I hear now that the lake is down over forty feet, and estimates are that they will be out of water in eighteen months if something isn't done. Other places—like Phoenix with water from an aquifer and two river systems—have a longer time for planning, but they all are under the gun. The heat's on!"

"Yeah," Sally chimed in, "and besides Mead's level falling fast, it continues to fill up with silt from upriver, although Lake Powell has relieved some of that. At the current rate, Mead will dry up in a few years if this drought doesn't let up and we don't conserve water more. But I know they are working on it. I hear desalination plants are coming to SoCal."

"That's certainly true," Bonnie added, "and we're happy to hear it because this river can't supply all the water for the region's growing population."

"At a point, I flew into Phoenix and took a drive up to the Canyon and then went back to I-40 at Flagstaff and drove over to Laughlin," Sally explained. "They have river taxis there that will take you from one casino to another. Lake Mohave north of the city is beautiful—one of several on the western side of Arizona. Zonies, as they're known in California, speak of their West Coast because of the reservoirs and communities along the Colorado—Parker, Lake Havasu City, and so on."

"Yes, we've traveled along the river there, too," Rob explained. "We also have been in Phoenix, and one time in our travels from there to SoCal we crossed the river at Blythe and on another occasion crossed it at Yuma. There's water in both places, but the flow is very slow in comparison, obviously, to the upper river where it has created all those canyons."

Then glancing at his watch, Rob exclaimed, "My word, it's nearly lunch time, and we have talked all morning. Not atypical of us retirees, I might add! We've had so much fun remembering our travels. We've been to so many of the same places along the river. Bonnie and I had better get back to our room. We have so enjoyed visiting with you. Perhaps we can again tomorrow?" Rob offered.

"Yes, I would like that very much," Sally responded as she accepted the invitation. "Perhaps we can meet for lunch." With that Rob offered his calling card to her. "I'll call you about a time," Sally offered. With that they parted company. By now the sun was high in the sky, and its rays had warmed the patio. As the couple once more entered the lodge and went into the lobby where they had been the night before, they gravitated toward the large plate-glass window framing that panoramic scene of the mountains.

"And to think that so much of that could be in jeopardy if we don't get more rain and snow," Rob whispered to himself. "Reduced precipitation would so hamper the Colorado's flow that millions of people, dependent upon it, could be affected. Major

Powell was right—it can't support millions of people indefinitely. Really, the heat's on—something must be done!" As Bonnie glanced at Rob and squeezed his arm, she took in the panoramic scene and, remembering the rugged land beyond this high mountain setting, sighed, "But, oh my, just think of what that river has created—such gorgeous *canyon country*!"

Made in the USA
Monee, IL
04 November 2026

67381823R10098